D1207370

Grazing in Temperate Ecosystems:
Large Herbivores and the Ecology of the
New Forest

GRAZING IN TEMPERATE ECOSYSTEMS LARGE HERBIVORES AND THE ECOLOGY OF THE NEW FOREST

R. J. PUTMAN
Department of Biology,
University of Southampton

CROOM HELM
London & Sydney

TIMBER PRESS
Portland, Oregon

© 1986 Roderick J. Putman
Croom Helm Ltd, Provident House, Burrell Row,
Beckenham, Kent BR3 1AT

Croom Helm Australia Pty Ltd, Suite 4, 6th Floor,
64-76 Kippax Street, Surry Hills, NSW 2010, Australia

British Library Cataloguing in Publication Data

Putman, Roderick J.
    Grazing in temperate ecosystems: large herbivores
    and the ecology of the New Forest.
    1. Ecology—England—New Forest
    2. Herbivora—England—New Forest
    I. Title
    574.5′264   QH138.N4
ISBN 0-7099-4036-X

First published in the USA 1986 by
Timber Press
9999 S.W. Wilshire
Portland OR 97225
USA

All rights reserved

ISBN 0-88192-071-1

Typeset in Plantin Light by Leaper & Gard Ltd., Bristol, England

**Printed and bound in Great Britain by
Biddles Ltd, Guildford and King's Lynn**

Contents

Preface

The New Forest in southern England is an area of mixed vegetation set aside as a Royal Hunting Forest in the eleventh century and since that time subjected to heavy grazing pressure from large herbivores. The entire structure of the Forest and its various communities has been developed under this continued history of heavy grazing, with the establishment of a series of vegetational systems unique within the whole of Europe. The effects of large herbivores in the structuring of this ecosystem in the past, and the pressure of grazing continuing to this day, have in turn a profound influence, indeed the dominating influence, on the whole ecological functioning of the Forest system. Because of its assemblage of unique vegetation types, the area is clearly of tremendous ecological interest in its own right. In addition, its long history of heavy grazing and the continued intense herbivore pressure make the New Forest an ideal study-site for evaluation of both short-term and long-term effects of grazing upon temperate ecosystems.

The New Forest (some 37,500 ha in total area) currently supports a population of approximately 2,500 wild deer (red, roe, sika and fallow); in addition 3,500 ponies and 2,000 domestic cattle are pastured on the Forest under Common Rights. From 1977, I have, together with a number of associates, undertaken a series of research studies on the ecology and behaviour of the large herbivores within the Forest, examining the various ways in which the different species use their Forest environment and considering their combined influence and impact upon the Forest vegetation. Slowly we are beginning to piece together some understanding of the complex functioning of this multi-species system, and to appreciate the influence of the heavy grazing pressure by the Forest herbivores on vegetational processes. Nor are the effects of grazing restricted to vegetational change: it is clear that, through its dominating effect upon the vegetation, the intense grazing pressure imposed by the larger herbivores has repercussions *throughout* the system — 'knock-on' effects upon other organisms reliant on this shared vegetational environment — until it affects in practice the whole ecological shape and functioning of the New Forest system.

In this book I have attempted to draw together the results of this work to present some kind of synthesis. It may be read as a series of separate studies within the New Forest of the ecology and behaviour of a number of different species of large ungulate: presenting a current review of our knowledge to date of the autecology of these species. I hope that it may also be read as a whole: as an investigation of the effects of heavy grazing on the dynamics and functioning of a temperate ecosystem.

A synthesis of this sort necessarily draws upon the work of many besides myself. It is a pleasure to acknowledge the debt I owe to all those whose work is quoted here, and to whom belongs the full credit for the tremendous contribution each has made to our understanding of the New Forest and its ecology. Particularly I would honour my research assistants Bob Pratt and Rue Ekins, who undertook most of the monumental amount of fieldwork that went into our studies of the Forest ponies and cattle, my various research colleagues and research students: Andy Parfitt (working on fallow deer), Chris Mann (sika deer), Steve Hill and Graham Hirons (whose work unravelled the complex effects of grazing upon the Forest rodent populations and their dependent predators) and Elaine Gill. I owe in addition an immense debt to my long-standing friend and colleague Peter Edwards, who has advised and assisted with supervision of the more botanical elements of our work throughout. Thanks are also due to various others who have worked within the Forest and have generously allowed me to quote their work: Norman Rand, John Jackson and Stephanie Tyler.

While I owe a tremendous debt to all these scientists, my thanks must also go to the Nature Conservancy Council, the Forestry Commission, the Verderers and all the New Forest Commoners who individually or collectively have supported our studies. Colin Tubbs of the Nature Conservancy Council first focused our attention on the New Forest system and persuaded the NCC to finance our initial studies of the cattle and ponies. His local knowledge of the Forest and its ecology is unrivalled and it has been a pleasure to have had his close interest and support in our work ever since. We owe a debt also to the various officers and staff of the New Forest Forestry Commission, who have over the years most patiently tolerated our tireless but frequently tiresome interest in the Forest. Far from merely putting up with us, they too have been active in support of our studies; we have made many friends in the

District Office and among the keepers and foresters. To all of these go my grateful thanks: I hope that this book, as the culmination of all those years of study, will contain something of use to them in the future management of the Forest.

Finally, I would thank Dawn Trenchard for coping so nobly with the horrendous task of creating an ordered typescript out of my scrawled chaos, Barry Lockyer and Raymond Cornick for help with the diagrams and other colleagues known and unknown who have commented on various bits of the manuscript itself.

Chapter 1

Introduction

Much attention has been devoted in recent years to the role played by large herbivores in shaping and maintaining vegetational systems. It is perhaps self-evident that in any system where grazing animals, whether large *or* small, occur in any density they are bound to exercise a significant influence upon that system: altering vegetational structure, diversity and productivity.

Although, typically, herbivores remove perhaps only in the region of 10% of the annual green-matter production in any community, in certain instances their impact may be far in excess of this. Wiegert and Evans (1967) estimate that ungulates may remove between 30% and 60% of primary production of East African grasslands, and Sinclair and Norton–Griffiths (1979) calculate that herbivores (both vertebrate and invertebrate) in the Serengeti National Park in Tanzania are removing up to 40% of the annual primary production. Even smaller herbivores can have a dramatic effect if they occur in sufficient density: in the tundra, arctic lemmings may remove up to 90% of available primary production (Schultz 1969), while even tiny invertebrates, massed together, can account for losses of between 10% and 30% of annual production of natural grasslands (Andrzejewska 1967; Schuster *et al.* 1971) or field crops (e.g. Bullen 1970, Clements 1978). Nor is the impact accurately measured by the amount of material actually ingested. Animal feeding is often wasteful: in studies of grasshoppers feeding on the black needle rush (*Juncus roemerianus*), Parsons and de la Cruz (1980) found that the insects consumed only 0.33% of annual net production; yet, as a result of the insects' habit of feeding in mid leaf, loss of production was three times that figure because of material clipped off and discarded. Red and roe deer browsing in deciduous forests in southern Poland show a comparable wastage: the deer remove some 46 kg (dry weight) of browse material per hectare; of this they actually consume

1

only 19 kg (Bobek *et al.* 1979), the balance being destroyed during feeding.

Further, it is clear that grazing may have a *far* greater impact on the community than is suggested by mere consideration of the absolute quantities of plant material removed: a herbivore ingesting perhaps 10% of a plant's production is going to have a far more significant effect upon the plant if that 10% is made up of primordia, destined for future growth, than if it merely results in a loss of 10% in the form of mature leaves. In the study of Bobek *et al.*, potential browse production of an unbrowsed forest was estimated as 172 kg/ha; actual production in practice (including the 46 kg later removed by deer) totalled only 160 kg/ha. Damage by deer *suppressed* productivity by 12 kg/ha, in addition to its effect in merely removing 46 kg of that production (Bobek *et al.* 1979). In another, similar study browsing by moose in a pine/mountain ash forest in Russia reduced forage biomass from 181 kg/ha to 109 kg/ha; only 3.5 kg of this loss was directly related to moose feeding, the remaining 68.5 kg being due to the reduced growth rate of the damaged trees (Dinesman 1967).

In fact, we may note a whole variety of these more subtle effects of grazers upon vegetation, where quite minimal *absolute* consumption by herbivores may have far-reaching significance.

(i) We may see a change in productivity. We have noted here a suppression of plant production due to herbivore pressure, but under other circumstances grazing may equally result in an *increase* in productivity.

(ii) Selective grazing by herbivores in a multi-species system may result in a change in species composition within the plant community as particularly graze-sensitive species are eliminated, and resistant, or tolerant, plant species may increase in dominance.

(iii) Grazing or browsing may alter natural patterns of nutrient cycling within the system.

(iv) Grazing may cause a change in the physical structure of the vegetation, altering the physical habitat and microclimate and making it more or less suitable for other plant and animal species, causing further changes in species composition — both within the plant assemblage and within the associated animal community (review by Putman 1985a).

The effects of herbivores upon *primary production* may be direct — through defoliation in feeding — or indirect: the trampling effects of hoofed mammals may lead to soil compaction and thus affect plant growth, while return of dung and urine to the system, resulting in local changes in soil-nutrient status, may also affect productivity. Direct removal of plant tissue by herbivores can directly affect rate of photosynthesis, respiration rate, location of nutrient storage, growth rates and phenology of the affected plant. While, as we have noted, heavy grazing and browsing may reduce production, moderate levels of grazing may actually increase plant productivity, through stimulating some compensatory growth. Growth will be inhibited — as in our example of deer browsing in Polish forests, where herbivores damage the growth primordia of the plant, or where excessive defoliation reduces the effective leaf area of the plant below a minimum threshold for efficient photosynthesis — but there are numerous examples in the literature where lighter grazing pressure can be shown to increase productivity. Productivity of natural meadowland in Russia was greatest at vole densities of 100 per hectare (Coupland 1979), when about 20% of plant growth was being harvested; in grazed sheep pastures in Australia, net production of a *Phalaris tuberosa/Trifolium repens* sward was greatest at a stocking density of 10 sheep per hectare (Vickery 1972). Similar increases in productivity in response to herbivory can be demonstrated for woody species. Various studies (e.g. Ellison 1960; Grant and Hunter 1966; Krefting *et al.* 1966; Wolff 1978) have established that trees and shrubs which are regularly browsed by ungulates frequently show increases in productivity under *light* herbivore pressure.

Such increase in production may be due to a number of factors. McNaughton (1979) lists nine possible mechanisms which may compensate for plant-tissue loss from herbivory and may result in increased production following grazing or browsing.

1. Increased photosynthetic rates in residual tissue.
2. Re-allocation of substrates from elsewhere in the plant.
3. Mechanical removal of older tissues functioning at less than a maximum photosynthetic level.
4. Consequent increased light intensities upon potentially more active underlying tissues.
5. Reduction of the rate of leaf senescence, thus prolonging the active photosynthetic period of residual tissue.
6. Hormonal redistributions promoting cell division and elonga-

tion and activation of remaining meristems, thus resulting in more rapid leaf growth and promotion of tillering.

7. Enhanced conservation of soil moisture by reduction of the transpiration surface and reduction of mesophyll resistance relative to stomatal resistance.

8. Nutrient recycling from dung and urine.

9. Finally, it has even been suggested that one potential stimulatory effect of grazing ruminants upon productivity may arise from plant-growth-promoting agents which have been found in ruminant saliva (Vittoria and Rendina 1960; Reardon *et al.* 1972, 1974). Direct growth stimulations up to 50% above control levels have been recorded following addition of ungulate saliva to surfaces of manually clipped leaves.

Note, however, that each plant species reacts differently to different grazing pressure — and to different grazers. While we suggested that maximum production of Australian sheep pasture was realised at a stocking density of 10 sheep per hectare, rabbits grazing on such pastures at densities as low as 40 per hectare can *reduce* pasture growth by up to 25%, merely because they feed in a different way on different parts of the plants; among woody species, too, treatments which best increase production of one shrub species may have different results for another species (review by Gessaman and MacMahon 1984). Defoliation by herbivores can thus stimulate *or depress* productivity of individual plants. Clearly, the actual result observed in any one instance will depend both on the degree of defoliation and also on the *timing* at which damage occurs in relation to the growth stage and growth characteristics of the particular plant. Different plant species will have markedly different responses depending on structure and growth pattern: the level of offtake at which productivity of woody species (with terminal growth points) begins to decline is far lower than that at which production of many grasses — which grow from the base — would be suppressed.

Where defoliation is sufficient to depress productivity, continued grazing may ultimately result in eradication of particular plant species from the community. Thus, through its effect upon the individual plant, grazing can alter the entire species composition of the community, with continued eradication of species sensitive to grazing and an increase in abundance of those species which, through chemical or physical defence, or because of their growth form, may have greater abilities to resist, tolerate or escape defoliation. Even without such deletion of species, however, defoliation may still have a profound effect on community composition. By reducing the leaf area of preferred forage species, the herbivore may reduce competition for light and space experienced by other plants, which may therefore be able to colonise or increase in abundance within communities from which they would normally be out-competed.

On upland grasslands in Britain, sheep show a distinct selection for palatable grasses such as *Agrostis* and *Festuca* species (the bents and fescues): these are grazed heavily in preference to the unpalatable *Nardus stricta* (mat-grass). Each year, therefore *Nardus* tends to increase at the expense of the other species (Chadwick 1960; Nicholson *et al.* 1970). Even the initial *Agrostis/Festuca* sward is itself a result of grazing. When sheep are excluded from such areas, the *Agrostis/Festuca* grasslands develop a patchy sward dominated by *Deschampsia cespitosa* (tufted hair-grass) and *Holcus mollis* (creeping soft-grass), and herbs such as lady's-mantle (*Alchemilla glabra*), sorrel (*Rumex acetosa*), *Galium* spp. and nettles (*Urtica dioica*) (Rawes 1981). (Interestingly, once *Nardus* has been allowed to become dominant within the sward, the change in species composition brought about by grazing is not so easily reversed. Long-term sheep exclosures on *Nardus* grasslands in the Pennines have shown little change in species composition in 24 years: Rawes 1981.)

Grazing of saltmarshes by sheep leads to the replacement of broadleaved species such as sea-purslane (*Halimione portulacoides*) and sea aster (*Aster tripolium*) by the saltmarsh-grass *Puccinellia maritima* (Gray and Scott 1977), and numerous other examples of changes in species composition in response to grazing may be cited. Thus, on semi-arid grasslands in the Serengeti National Park in Tanzania, McNaughton (1979b) records marked differences in species composition of grazed areas from that observed within plots protected by fences since 1963. In one comparison he noted that the grasses *Andropogon greenwayi* and *Sporobolus marginatus* made up 56% and 20% respectively of biomass outside the exclosure, but were not recorded in samples taken inside; in contrast, *Pennisetum stramineum* and *P. mezianum* made up only 5% and 3% of the biomass in grazed areas, while contributing to the biomass of ungrazed plots by 72% and 26% respectively. Perhaps the oldest and most quoted example of how grazers can affect the species composition of whole vegetational communities is in the development of the specific plant assembly we now associate with chalk grassland in Britain, under the influence of first sheep, and later rabbit grazing (Tansley 1922;

Tansley and Adamson 1925; Hope Simpson 1940).

Grazing herbivores not only affect productivity and species composition of the vegetational communities on which they graze: they may also have considerable impact on nutrient cycling (which may in itself have yet further repercussions for species composition and productivity of the vegetation: McNaughton 1979, Crawley 1983). By feeding on plant materials, animal lock up within their own tissues essential nutrients, making them unavailable to the next 'generation' of plants. In systems where nutrients are relatively abundant, and there exists in the soil a relatively large pool of 'free' nutrients, this has but little effect upon plant growth; but in other, nutrient-poor systems, the effects of having significant quantities of a limited nutrient supply bound up in animal tissue may well be quite marked. In the tundra zone of Alaska, for example, lemmings, in years of peak abundance, may consume 40% of the available nitrogen and 50% of the phosphorus. Although ultimately all of this is returned to the soil-nutrient pool in the form of animal wastes (in faeces or as carcasses), there is a considerable time delay, and for some period the vegetation becomes markedly nutrient-impoverished. The situation is complicated in this example by the cyclic nature of lemming population abundance. When lemming populations are low, few of the available nutrients are tied up in animal tissue, so plants have adequate nutrients available to them and grow well. Offered such abundant and nutrient-rich vegetation, lemming populations grow fast and reproduce rapidly. As lemming numbers increase, so fewer and fewer nutrients are left available for plant growth. Productivity and quality of the vegetation falls, which leads in turn to a decline in the reproductive performance of the lemmings themselves. At this stage in the cycle, natural mortality outstrips reproductive recruitment: reproductive rates are low, and those individuals born into the population in the last boom are reaching the end of their allotted span. Lemming numbers fall; decomposers release the nutrients of their body tissues, which once more become available to the vegetation — and the cycle begins again (Schultz 1964, 1969).

While such effects are really important only in nutrient-poor systems, *patterns* of nutrient cycling may be affected by grazing in any system. Animals that feed over a wide area but defecate in a small area can have a substantial effect on local nutrient distribution. Sheep, for example, graze widely over pasture during daylight, but congregate in camps at night or for shade; in consequence, 35% of their faeces are deposited on less than 5% of the grazing area — resulting in a gradual impoverishment of the wider grazing range, but continued enrichment of small areas within it (nutrient '*dislocation*': Spedding 1971). Other animals may show different habitat preferences for grazing and for elimination, so that nutrients may be removed from some habitats and returned to others; this *translocation* of nutrients, too, has a profound effect on the nutrient dynamics of any community. While these are effects of elimination more than grazing, pure defoliation itself may alter the water balance of the community, reducing the area of transpiring leaves, and exposing areas of the soil surface to the drying influence of sun and wind (Crawley 1983).

Finally, as we have noted, although most work has concentrated on the *direct* effects of grazers upon the vegetation itself, it is clear that the effects of the herbivores are not restricted to this level, but have a whole series of 'knock-on' effects. All the various effects of grazing and elimination result in modification of the habitat itself — and also of the environment it offers to others. A changing microclimate, through structural modification of the primary vegetation, will have an effect on the secondary plant species, and animals, which may colonise the 'new' environment. Changing species composition, and species dominance, will affect relative availability of food to other smaller herbivores. In short, through the changes that the grazing process causes within the vegetation — in structure, species composition and productivity — it has at once additional effects on the rest of the community dependent on that vegetation.

In conclusion, the potential effects of heavy grazing upon vegetation may be summarised as: increase or decrease in primary production; alteration to nutrient cycling and gross distribution of nutrients; changes in species composition (due in turn to the selective elimination of sensitive species, to changes in the competitive equilibrium of the plant community under grazing pressure, and to modification of nutrient flows within the system); and changes in the actual physical structure of the vegetation itself, affecting microhabitats offered to dependent plant and animal species.

There is yet one further whole category of animal influences on plants, where the consumer affects *other consumers* which in turn affect the plants. Gessaman and MacMahon (1984) cite the example of a predatory sea-otter (*Enhydra lutris*) feeding on invertebrate herbivores such as abalones and sea urchins (Simenstad *et al.* 1978). The removal of the invertebrate grazers increases the density and biomass of macro-algae. But predation is by no means the only biotic interaction between consumers that may alter their impact upon vegetation: competition between the herbivores themselves, in multi-grazer systems, may influence their individual effects upon the vegetation. Illustration of each of these effects may be drawn from examination of the multi-species grazing system found in the New Forest in Hampshire — a system which also provides us with a unique opportunity to investigate the overall combined effect when all these separate, potential influences act together within a single ecological system.

The New Forest

The New Forest in southern England is an area of mixed vegetation set aside as a Royal Hunting Forest in the eleventh century and since that time subjected to heavy grazing pressure from a variety of large herbivores. The entire structure of the Forest and its various communities has been developed under this continued history of 900 years of heavy grazing. Many of the Forest's vegetational communities owe their very existence to past and present patterns of grazing, and the area boasts a number of particular vegetational formations now unique within Europe. Such a system offers an unparalleled opportunity to study in detail both long-term and short-term effects of intensive grazing within a single system, while the current ecology of the area itself can only be interpreted in relation to the various effects of past and present grazing. Any attempt to describe the ecology of the Forest — in accounting for the curious lack of diversity of many vegetational systems, low numbers and diversity of small mammals, curious behaviour of birds of prey and other predators, unusual patterns of habitat use of the larger herbivores themselves — any attempt to *explain* rather than just describe, forces the attention back to the dominating effect of grazing in the shaping of this ecosystem. An account of the 'local' ecology of the New Forest, and an examination in more general terms of the ecological effects of heavy grazing, thus prove quite inseparable.

Set in the heart of the Hampshire Basin and covering an area of some 37,500 ha between the Solent and the Avon (Figure 1.1), the New Forest is more easy to define as an administrative area than as an entity. It is a diffuse region of Common land, dissected by roads and railways, perforated by towns and quite sizable agricultural holdings which have grown up within its boundaries on the better soils.

Figure 1.1: Location of the New Forest in the South of England

Even the term Forest is somewhat misleading, suggesting as it does an area densely wooded, for in fact only some 10,000 ha are actually covered with trees. In common with most royal forests, the word 'Forest' is in this context used in its older sense of an area set aside for hunting game (the modern association of a forest with trees is a purely secondary association), and in practice much of the area is open heathland or grassland.

The diversity of vegetation types within the Forest is in fact far greater than might be expected for the area, and this may be attributed to considerable variations in edaphic factors over the area. Heavily leached and base-poor plateau gravels and sands are widespread, particularly in the north, and support a *Calluna*-dominated *dry-heath* community. At lower altitude, and where the plateau gravel has been eroded, more fertile clays and loams support a mixed *deciduous woodland.* These are predominantly of beech and oak, with an understorey of holly; the common bent (*Agrostis tenuis*) colonises the woodland floors, in openings and glades. (Many of the more fertile woodland sites have been enclosed over the last 100 years and now support commercial plantations.) Also common on the more fertile soils are a range of *acid-grassland* communities, dominated by the coarse grass *Agrostis setacea* and to a lesser extent the purple moor-grass (*Molinia caerulea*), usually colonised to a greater or lesser degree by bracken and, particularly in the south, also by gorse (*Ulex europaeus*).

Where drainage is impeded in the valley bottoms, domination of the heathland by *Calluna* is diminished, and the species diversity of the whole heath community increases. A gradation is observed from the dry-heath community through *humid* and *wet heath,* with increasing abundance of cross-leaved heath (*Erica tetralix*) and *Molinia* and the appearance of true wetland plants such as bog asphodel (*Narthecium ossifragum*) and *Juncus* species. This progression frequently ends in bog communities. The *valley bogs* offer some of the richest vegetation types in the New Forest in terms of plant diversity, and are one of the formations unique to this area. The species composition varies considerably in relation to how eutrophic the water supply is, and several distinct communities can be recognised. Perhaps the most widespread in base-poor water is that dominated by *Molinia* tussocks with common cottongrass (*Eriophorum angustifolium*) and *Sphagnum* mosses abundant between the tussocks. In many heathland catchments, *carr woodland* develops in the valley bottom where drainage waters have a definite axis of flow. These carr woodlands are composed of *Salix atrocinerea,* alder buckthorn (*Frangula alnus*), alder (*Alnus glutinosa*) and other tree species, and have a diverse herb layer including the greater tussock-sedge (*Carex paniculata*).

Where the area is well drained by one of the many small streams which transect the Forest, the bogs are replaced and the heathland progression terminates abruptly at the edge of alluvial strips bordering the streams. These alluvial deposits are covered by grassland, often dominated by *Agrostis canina,* interrupted with patches of riverine woodland. These *streamside lawns* are particularly nutrient-rich because of regular annual flooding from the rivers which they border, which carry base-rich compounds from north of the Forest.

Very little of the New Forest vegetation can be considered natural, and most areas have at various times been subjected to management by man. This has resulted in the creation of a number of new community types, the most obvious of which are the *re-seeded lawns*, areas of acid grassland which were fenced during the Second World War, ploughed, fertilised, and cropped for potatoes and oats. At the end of the war, these were re-seeded with a ley and after the grass-land was established the fences were removed. Many of the sown species have since disappeared from the sward, which has been recolonised by natural grassland species, but the areas still comprise a very distinct vegetation type. In the late 1960s and early 1970s, other attempts to improve the grazing and grassland were made by swiping bracken from certain areas and liming the grassland. These improved areas once again form a distinct and characteristic vegetational community. Finally, the heathland communities are subjected to con-tinued management, being cut, or burnt on a rotational basis, so that any extensive area of heath contains a patchwork of sub-communities from 0 to 12 years of maturity.

Of the total administrative area of the Forest, some 9,000 ha are occupied by towns, villages or agricultural land; 8,300 ha have been enclosed for commercial plantation; while only some 20,000 ha remain as 'wasteland' or 'Open Forest' whose vegetation we have just described. This 20,000 ha, with its mixture of vegetational com-munities, supports a big population of large herbivores. Some 2,500 wild deer (red, roe, sika and fallow) have access to the entire area, while the open, unenclosed Forest supports in addition approximately 3,500 ponies and 2,000 cattle, pastured on the Forest under ancient Rights of Common.

Of the four deer species present on the Forest today, red deer (*Cervus elaphus*) and sika (*Cervus nippon*) are essentially local in distribution: populations are restricted to limited areas of the Forest. Roe (*Capreolus capreolus*) are everywhere uncommon, and only fallow deer (*Dama dama*) are both widespread and abundant (num-bers estimated at about 2,000 in 1976: Strange 1976). (It has been suggested that muntjac deer, *Muntiacus reevesi*, may also now be resi-dent within the Forest woodlands, but reports are unconfirmed and numbers are in any case low.) Red deer and roe are both native to Great Britain. Fallow deer were introduced by the Romans or Nor-mans and may be considered an old-established British species; sika have been introduced more recently. Within the New Forest, populations of both fallow and roe are natural, and must be presumed to have existed since the Forest's creation in 1069. Red deer were also almost certainly native to the area, but present-day populations derive from two separate introductions in different parts of the Forest (page 112).

Domestic animals are also grazed upon the 20,000 ha of the Open Forest. One of the concessions granted to the local populace when the area was declared a Royal Hunting Forest was the right of common grazing. On the right payment of an appropriate 'marking fee' (presently set at £10 an animal), local cottagers and farmers could turn out cattle and horses to exploit the rough grazing of the forest lands. These rights are still honoured, and in 1980 marking fees were paid for some 3,400 ponies and 2,000 cattle (although this of course does not mean that all these animals were necessarily on the Forest all through the year).

In times past, the Forest deer were probably by far the most significant grazing pressure upon the New Forest vegetation: the area was, after all, set aside for the preservation of game and, until the 1850s, deer populations over the area numbered between 8,000 and 9,000. At the end of the nineteenth century, however, the area was 'disafforested'; deer populations were decimated and have only recently recovered to their current status. Their present, lowered, density now casts the free-ranging domestic stock as the major herbivores. Cattle and ponies have been depastured on the Forest alongside the deer ever since its designation as a Royal Hunting Preserve in the eleventh century, and probably considerably before that time, but numbers were probably lower than at present and in addition a far larger area of land was unenclosed and available for common grazing. As a result, the impact of the common stock upon the Forest was probably secondary to that of the deer. With reduction in the numbers of deer and simultaneous increased effective density of domestic stock at the end of the nineteenth century, however, cattle and ponies emerged as the major grazing influence, and have continued so to this day.

The New Forest area has, then, always sustained a tremendous grazing pressure from large herbivores. At present, 20,000 ha of some of the poorest possible grazing (current land-use surveys class the majority of the area as grade 5, or non-agricultural land) support some 8,000 head — a total biomass in excess of 2,500 tonnes — and it is clear that equivalent grazing pressure must have existed over the centuries. Yet the system has maintained itself in some kind of balance, however uneasy, for more than 900 years, this without any formal attempt at long-term management. Indeed, the system has survived *despite* dramatic and drastic changes in management and land-use (Chapter 2). This history of continued grazing has, however, stamped its mark upon the Forest vegetation, and provided a major influence on the shaping of the Forest and on its present-day ecology.

Over the last 40-50 years the Forest has come under increasing pressure, from commercial interests, tourism and recreation. Rising beef and dairy prices and the establishment of a ready market for horse flesh have produced a steady increase in the numbers of domestic stock commoned on the Forest since the war; commercial forestry interests in the area continue, and the Forest has also become more widely used as a recreational centre, for camping, walking etc. In order to maintain the Forest for the future, to conserve all its varied resources for its many users, and to conserve its unique vegetational structure, some deliberate policy of management must now be adopted. To be of value, this must itself be based on a firm understanding of the ecological functioning of the New Forest system, past and present. To this end, the Nature Conservancy Council commissioned a study in 1977 of the current interrelationship between the grazing herbivores of the New Forest and the Forest vegetation. Initially restricted to a study of the ecology of the Common animals — cattle and ponies — as the most significant of the herbivores, its aims were to discover the pattern of use by the cattle and ponies of the different vegetation types available to them within the Forest and also to estimate the impact of that pattern of habitat use and feeding activity upon the vegetational community in calculating the effects of grazing pressure on vegetational productivity and diversity.

The deeper we delved into the ecology of the Forest animals, the more fascinating became the complexity of interaction between animals and vegetation — and animal and animal. Over the years our studies ranged wider, to include consideration of the current role of the Forest deer in this complex ecological community, and to consider the effects of such heavy grazing on other parts of the ecosystem. Grazing proved to have such a dominating effect on vegetational structure and species composition that it became clear that the effects of the grazers on the vegetation must also have considerable repercussions on the ecology of other organisms dependent on the same vegetation; we therefore extended our studies to consider the effects of large herbivore pressure on the smaller herbivorous mammals of the Forest, and, through them, on their various predators. This book aims to bring together the results of all these various studies, offering a comprehensive review of our accumulating knowledge of the autecology of the various animal species of the Forest. Through our apprecation of the interactions between these species and between them and the Forest vegetation, it attempts then to draw the separate threads together into an integrated whole, to produce a new insight into the actual ecological processes underpinning the Forest we observe. In this sense the book builds naturally on Colin Tubbs' classic 'ecological history' of the New Forest (Tubbs 1968), in updating our information in the light of recent work and in presenting a picture of the ecological functioning of the Forest as it is today.

The ecological functioning of the New Forest is so dominated by the effects of heavy grazing that the book has in addition a secondary interest: as a detailed analysis of the effects of continued heavy grazing on a temperate ecosystem. Although the focus of the book remains an analysis of the ecological functioning of one complex, multi-species ecosystem, our studies in the New Forest have important implications in terms of the effects of grazing on vegetation in general. We shall hope therefore to provide, through this 'case study', insight into the more general effects of grazing in the dynamics of such temperate ecosystems, in terms of those same, potential effects outlined in the first few pages of this chapter.

Chapter 2
The History of the Forest

Despite the fact that it may claim to be the largest single unit of 'unsown' vegetation remaining in lowland Britain (Tubbs 1968), the New Forest is none the less a man-made system. The shaping of the Forest owes as much to its past history of human management as to its current grazers. The two are, of course, not entirely independent. Man's activities clearly affect relative numbers of wild herbivores within the Forest area today, as in the past. Man's domestic animals, too, are major contributors to overall levels of grazing within the Forest. But other human activities have also affected the Forest and its development, and continue to do so. The whole development of the Forest may be seen as a chronicle of human activity and human pattern of land-use — and Nature's continued response to the different pressures of each generation. The current ecological functioning of the Forest system is influenced by this history and by current methods of land-use; these modern activities are themselves restricted by the laws and traditions of the Forest, and by an administrative system which has evolved over 900 years. The Forest today is a complex product of *all* such factors, human and biological. Fully to appreciate the ecology of this curious system it is therefore critical to understand its social context, both historical and present-day.

The Early History of the Forest

Regrettably, very little is known of the history of the New Forest area before its annexation as a Royal Forest in the eleventh century. Using what little evidence there is and a great deal of deduction, it is, however, possible to piece together a picture of land-use. Mesolithic sites on heathland are widespread, and are the first good evidence for extensive use of the area. Mesolithic peoples, however, left few

15

obvious traces on the ground: they did not cultivate the land and lived essentially as hunters. Their major impact on the Forest system was probably in the clearance of trees. The rapid regeneration of ground vegetation and shrubs in the secondary succession of forest clearings provides, in the short term, a greatly increased supply of rich forage for wild game. Forest clearance in provision of this increased grazing has always been a part of the human hunter-gatherer economy, and Mesolithic man was no exception. Large-scale clearance of forest and forest edge, usually by fire, was quite common in this period; other areas may have been accidentally cleared by fires lit to drive or flush game.

While Neolithic and Mesolithic peoples left few traces on the landscape, the succeeding Bronze Age cultures left more tangible evidence of their occupation in the form of burial mounds or round barrows. In both the Middle and Late Bronze Age, there is therefore abundant evidence for human occupation of heathlands within the Forest. Pollen analysis of soils buried beneath some of the barrows suggests that, at least by the Middle Bronze Age, some form of actual cultivation of the heathlands was being practised. This cultivation was short-lived, however, and apparently did not persist into the Iron Age; indeed, evidence for Iron Age occupation of the Forest heathlands is in any case very sparse. Tubbs (1968) suggests that this scant evidence of Iron Age occupation, by comparison with the abundant evidence for Bronze Age settlement, reflects not just an archaeological vacuum, but actual desertion of the Forest heathlands because their fertility had declined to a point where cultivation was no longer possible.

The forest clearance of Neolithic and Mesolithic times probably continued through the Bronze Age, though its purpose may have shifted towards the creation of clearings for agriculture. In the short term forest clearance may encourage rapid growth of lush grazing for wild or domestic stock; it may provide relatively fertile areas for cultivation. But this apparent productivity is short-lived: on poor soils like those of the New Forest, clearance of forest ultimately leads to a dramatic deterioration in soil fertility as changed rainfall and run-off patterns lead to leaching of the limited soil nutrients. Thus Bronze Age peoples probably settled in areas which were already beginning to decline in fertility; by the Iron Age, the soils were in general no longer suitable for cultivation by the techniques then available.

After the end of the Bronze Age therefore, and certainly by the end of the Iron Age, the open heathland areas of the Forest were exhausted of nutrients, and usable only for rough grazing. Instead of their previous widespread distribution across all the Forest lands, settlements became restricted only to those areas of better soil which could still support cultivation. A pattern of settlement became established which persisted through Saxon and Norman times, and which indeed has largely persisted until the present day, with communities focused upon the limited number of areas where soils are rich enough to support arable agriculture, and the greater part of the Open Forest deserted except for exploitation of its limited grazing.

The Royal Forest

Undoubtedly, the most significant event in the history of the Forest was the appropriation of the area in 1069 by the Crown as a Royal Forest. This dramatically affected its ecological development, in markedly influencing subsequent patterns of land-use. It has also, to an extent, safeguarded the Forest and its peculiar mixture of communities. Although by the middle of the eighteenth century agricultural expertise had developed to a point where reclamation of even the most infertile soils for agriculture was a practicality, the Forest and its open wastes, as Crown land, were safe from reclamation.

As noted in the previous chapter, designation of an area as a Royal Forest implied establishment of a hunting preserve for the Crown. It did not necessarily imply that an area was wooded, nor indeed did it necessarily imply Crown ownership of land. Afforestation was based on the prerogative enjoyed by the Sovereign that all wild animals were in his possession. The later Saxon kings claimed as part of this prerogative the right to reserve to themselves the chase, at least of deer, over any part of their kingdom they might choose to define; under the Norman kings, the restrictions of the use of these royal hunting grounds became even more severe and were established in statute. The King had, in effect, the right to subject any land, whether of his ownership or owned by others, to Forest Law: a forest, while it has a secondarily assumed implication of woodland cover, was thus defined in fact simply as an area coming under Forest Law as opposed to Common Law. While imposition by the Crown of Forest Law did not imply any legal change in ownership, it none the less brought with it severe restrictions: lands under Forest Law could be neither enclosed nor cultivated; the rights to take game were reserved for the Crown. The restrictions were such that the private owner of such land gained little from that ownership; in 'compensation' he was allowed to retain the rights of free-grazing of stock over the area. Clearly, it would have been unrealistic to prohibit fencing and at the same time to enforce restrictions on the roaming of stock from private land on to Crown land. Thus the right of free range of stock over the entire area seems to have gained mutual acceptance (and would appear to be the origin of many later rights of Common grazing on Crown lands, although these were not legally granted on *any* Crown land until very much later, in 1598).

The New Forest was designated a Royal Forest by William I, and of all the many Royal Forests created during the eleventh, twelfth and thirteenth centuries it alone persists today in near entirety. The form of management imposed by Forest Law created a pattern of land-use which shaped the whole development of the area, and which remains much the same to the present day. Although subsequent reforms progressively sweetened the old Forest Laws, they did not fundamentally change their import. (As a result of growing disaffection among landowners for the restrictions of Forest Law, the *Charta de Foresta* of 1217 conceded extensive disafforestation and apparent relaxation of restrictions. Before the *Charta* no land under Forest Law could be enclosed or cultivated, nor could timber be felled or game taken. Under the *Charta* these practices were permitted on private land, but only under special licence from the Crown. In effect, the old Forest Laws remained in force right up until the New Forest Deer Removal Act of 1851, when the Crown relinquished its right to keep deer in the New Forest.) But, while the administration of the Forest changed little over the next few centuries, its boundaries did. Increasing demands on land, and increasing land values, put tremendous pressure on the Crown's hold of the Forest preserves. Neighbouring estates encroached on the outer boundaries. Richer lands within the perambulation were enclosed for cultivation, either under licence from the Crown or in flagrant disregard of the Forest Law. The initial area of lands afforested by the Conqueror had diminished to some 67,000 acres (27,110 ha) by 1279, when a firm perambulation or legal boundary was established. But, while these same pressures resulted in the gradual fragmentation or total dissolution of other Royal Forests, the New Forest somehow survived virtually intact: the perambulations of 1279 remained more or less unchanged until recast by the New Forest Act of 1964.

Patterns of land-use also changed, even on the open Forest. While the original purpose of the Forest was the preservation of deer, the value of the area for timber production became more and more important as England emerged as a seafaring nation, and as the country's total timber resources began to dwindle as they were rapidly consumed in the shipyards. From the seventeenth century onwards, the Crown's interests in the New Forest were focused more and more on the production of timber. Tree production on a commercial scale, however, required the enclosure of land against browsing and grazing by deer and commonable stock. More and more land within the Forest was enclosed — both by the Crown and by private landowners — for silviculture, culminating in the Enclosure Acts of 1698 and 1808. This new emphasis on timber production and the resultant enclosures began to change the character of the Forest and was the next major stage in its development. As more and more land was enclosed for silviculture, less remained available for Common grazing. Tubbs (1968) notes that the documentary evidence of the seventeenth, eighteenth and nineteenth centuries shows that a vigorous pastoral economy based on the use of these Common grazings had developed in the Forest. While numbers of animals depastured on the Forest doubtless decreased as land was enclosed, actual densities of animals on the remaining acres of open ground must thus have increased tremendously, leading to extremely heavy grazing pressure on these areas — and another major 'event' in the shaping of the Open Forest's characteristic vegetation patterns.

This same conflict of interests between those of the Crown (in enclosing land for silviculture) and those of the Commoners (in retention of the open grazings) culminated in the New Forest Deer Removal Act of 1851, by which, as noted earlier, the Crown relinquished its right to keep deer in the New Forest and released the area from many of the restrictions of Forest Law. Under the same law, the Crown took for itself instead powers for silvicultural enclosures on an enormous scale — well beyond the total 6,000 acres (2,430 ha) provided for by the Enclosure Acts of 1698 and 1808. Twenty-five years later, in the New Forest Act of 1877, the powers of the Crown to enclose land were, however, severely curtailed and the Commoners were assigned a statutory area of 45,000 acres (18,210 ha) over which they might exercise their rights in perpetuity. But, from this time on, forestry — in its sense of timber management — has been an important part of the New Forest economy, and an important influence on the vegetational structure of the Forest. There are currently some 20,000 acres (8,300 ha) of statutory silvicultural enclosures within the Forest, although only 17,600 acres (7,100 ha) may be behind fences at any one time.

Management of the Crown-owned lands of the New Forest passed to the Forestry Commission in 1923 — with responsibility for timber production within the Statutory Inclosures, and maintenance of Common grazing. At the same time the Forest has come under increased pressure for recreation. Management of the area, both Inclosures and Open Forest, must thus be framed to preserve in addition the amenity value of the Forest; indeed, the Commission now has a statutory requirement so to do. In our century, then, as in all the years before, the Forest must serve a multitude of purposes. Today its management is for timber production, for maintenance of Common grazing and Common agriculture, for recreation and for conservation. These management aims will play as important a part in shaping its future as its past history has had in shaping its present.

The Forest Administration

During the thirteenth century, Forest Law was administered by two main courts. The lower court, the Court of Swainmote or Attachment, was presided over by appointed Verderers of each Forest. Its function was essentially to hold preliminary hearings, binding over those prosecuted by Forest Officers for breach of the law to the higher court, namely the Justice in Eyre. The Court of Swainmote, or the Verderers' Court, had no authority itself to impose fines or punishments. The Justice in Eyre was supposed to hold court in each Forest every three years, but by the end of the thirteenth century it was held only irregularly; in the seventeenth century only two courts were held within the New Forest, the second, in 1670, being the last court ever held. This gradual demise of the Justice in Eyre was doubtless in large part due to the progressive relaxation of Forest Law, softened from the original toughness of Norman times by successive Charters. But without a Supreme Court to which to refer, the Court of Swainmote and Attachment became to a large extent powerless in any legal sense, and began to function more as a manorial court, merely administering privileges and rights of both Crown and Commoner within the Forest itself.

The executive officer in charge of a Forest was the Warden or

Lord Warden, a post which could be hereditary or by Royal Appointment; the Verderers of the Forest were elected by the County. Day-to-day management and policing of the Forest were carried out by Foresters and other minor officials. Agisters were employed by the Verderers to look after stock on the Forest and to enforce the by-laws, and the Regarders (originally 12 knights of the county) were responsible for examining the Forest perambulations and reporting on the general state of the Forest.

In the New Forest, the Verderers' Court still remains; the new role as protector of Common Rights, adopted by the Court of Swainmote and Attachment after the collapse of the Justice in Eyre, was confirmed and given statutory definition by the New Forest Act of 1877. This Act redefined the Verderers as a statutory body responsible not for safeguarding the interests of the Crown (for which the Forest Courts had originally been intended), but for safeguarding instead the interests and rights of the Commoners. The New Forest Act of 1949 reconstituted the Verderers for a second time, increased their number from six to ten and extended their powers. The Court now consists of the Official Verderer, nominated by the Crown; five Verderers elected by persons occupying an acre or more of ground to which Common Rights attach; and four Verderers (added by the 1949 Act) appointed respectively by the Ministry of Agriculture, the Forestry Commission, the Local Planning Authority and the Council for the Preservation of Rural England. The Verderers' responsibilities are to make by-laws for the control of stock on the Forest and for the benefit of the health of the Commonners' animals; to employ officers as required for the control and management of commonable stock (the Agisters); to defray costs by levying dues ('marking fees') for each animal turned out on the Forest; and to maintain the Register of Common Rights.

In brief, the responsibilities of the Verderers remain those of defending and safeguarding the interests of the Commoners. Common pasturage, however, is only one facet of the current economy of the New Forest. As we have noted, it is still a Royal Forest where the Crown retains silvicultural interests. In effect, the ancient and perpetual conflict of interests between Crown and Commoner which has for so long been a part of Forest history continues to the present day, with the modern-day interests of the Commoners represented by the Verderers and the interests of the Crown represented by the Forestry Commission (to whom exercise of the Crown's interests has been devolved). While the conflict between the two *bodies* has to a large extent been overcome in recent years (the Forestry Commission is represented on the Court of Verderers, while the Verderers are invited to join in discussions about Commission management policy), the conflict of *interest* between silviculture and Common pasturage is a real one.

The *administration* of the Forest is thus the responsibility of two statutory bodies (Commission and Verderers), in opposition or in collaboration. The actual *management* of the Forest is equally torn by conflict of interests: to the interests of Commoners in grazing, and of Crown in timber, are added statutory requirements for conservation (whose interests are represented in the New Forest by the local officer of a third body, the Nature Conservancy Council) and for provision of recreational amenities. The New Forest Act of 1964 adds to the responsibilities of the Forestry Commission and Verderers the obligation to 'have regard for the desirability of conserving flora, fauna and geological and physiographic features of special interest' within the Forest; the interests of amenity are safeguarded in earlier Acts, particularly in regard to the preservation of the ancient woodland of the Open Forest (the so-called Ancient and Ornamental Woodlands of the Open Forest were secured under the 1949 Act).

The current administration of the Forest must thus try to satisfy four major aims, often in conflict; and that administration is the responsibility of *two* statutory bodies, in consultation with other interested parties such as the Nature Conservancy Council and the Countryside Commission.

Rights of Common

While the history of changing patterns of land-use within the Forest has clearly had a profound influence on its ecological development — and while its current complex administration determines the direction of overall management policy in the future — these are long-term and indirect influences. On a day-to-day basis the ecology of the Forest was, and is, affected *directly* by the exercise of various Rights of Common, and it seems appropriate to end this section with a brief review of these Forest rights.

Before afforestation in 1069 the open wastes of the Forest would have been freely grazed. Pigs would have been turned out to feed on acorn and beech mast in the autumn; timber, turves or peat taken for fuel; and bracken cut for bedding and litter. After the imposition of

Forest Law, these practices were allowed to continue, but under close control and regulation — eventually, within the New Forest, being defined in law in the Acts of 1689. These rights may be summarised as Rights of Common of pasture for commonable beasts (cattle, ponies, donkeys); Right of Common of mast (the right to turn out pigs during the mast season, or 'pannage'); Right of Common of turbary (the right to take turf fuel); Right of Estovers (the right to take fuel wood); and Right of Common of marl (the right to take marl from recognised pits in the Open Forest). To these may be added rights to cut 'fern' (bracken) for bedding and litter, and sheep rights claimed by the adjoining manors of Beaulieu and Cadland.

Common Rights could be claimed by any cottager or landowner holding more than an acre of land within the Forest perambulation. While most claims to pasturage conceded the old rules under Forest Law of *levancy* and *couchancy* (limitation of the number of stock turned out in spring and summer to that number which could be maintained on the holding over winter), this was clearly not a rigid requirement (nor is it today) and many claims are for the right of pasturage in all months.

Both cattle pasturage and pannage of pigs were, however, subject to certain restrictions under Forest Law. Medieval Forest Law provided for the removal of cattle from the Royal Forests during the midsummer 'fence month' (20 June to 20 July) when the deer were calving; and during that part of the year when keep was short (22 November to 4 May) — the winter 'heyning'. It would appear that the fence month was observed for the most part within the New Forest, but it is clear from the various Registers of Claims recorded from 1635 onwards that the winter heyning was largely neglected. Under Forest Law, the period during which pigs might be turned out to mast was restricted to about two months in the autumn (the precise dates were set each year, according to the mast-fall). Yet, once again, while presumably rigorously observed in the early days when the area first came under Forest Law, these restrictions were but weakly enforced by the sixteenth and seventeenth centuries and in the first detailed Register of Claims compiled by the Regarders in 1635 many claims for Right of Common of mast are not confined to the true pannage season: many are claimed for all times of year.

Rights of pannage and Common pasturage were undoubtedly the most significant and important rights which could be claimed by Forest Commoners in terms of maintaining a livelihood; it is clear that many of the holdings were of such a size that they would not have been viable without these rights upon the Forest. In evidence given before Government Select Committees in 1868 and 1875, it was stated that the Right of Common of grazing enabled a Commoner to maintain at least three times as many cattle as he would have been able to maintain without that right. But almost all claims registered in 1635 are for every Right of Common, and there is no doubt that the rights to take bracken for bedding and litter and, more particularly, the rights to take turf and fuel wood must have contributed significantly to the cottage economy.

While Registers of Claims to Common Rights on the New Forest have been compiled at regular intervals since the seventeenth century, such records cannot show the extent to which these rights were exercised. Thus it is difficult to assess the impact that the exercise of the different Rights of Common might have had upon the Forest at different times. Precise figures for the number of stock depastured on the Forest are available for the years 1875, 1884-93 and from 1910 onwards. A census of stock carried out in 1895 gave a figure of 2,903 ponies, 2,220 cattle and 438 sheep. The number of stock for which marking fees have been paid has fluctuated considerably since that time — and presumably did before it — depending largely upon the outside economic situation. By 1910 the number of ponies on the Forest had fallen to 1,500; cattle numbers remained between 2,000 and 2,500. In 1920 numbers rose again to a peak of 4,550 stock (mostly cattle) before the depression led to a slump in Common practice, and on the outbreak of war in 1949 there were only 1,757 domestic animals on the Forest. Economic ups and downs are sensitively reflected by ups and downs in the numbers of large stock turned out to Common: numbers have fluctuated between 1,500 and 5,500 over the last 100 years (Table 2.1). These same differences in the degree of exercise of Common Rights have had profound effects upon patterns of vegetational change within the Forest.

The commonable stock and other large herbivores have grazed upon the Forest vegetation for centuries, and have shaped and modified the structure and function of the entire Forest community. Whatever effects they may have upon the Forest's ecology they have had for hundreds of years, yet remarkably little is known of the animals' day-to-day use of the Forest in which they live and which they have helped to shape. Their alleged role in the dynamics of the Forest system past and present is largely supposition, based on anecdote and circumstantial evidence. It is of course not now possible to

Table 2.1: Commoning activity in five periods of time since 1789

Historical period	Number of Commoners	Number practising		Number of ponies	Number of cattle
1789-1858	2,314	800 to 1,000	MAX	3,000	6,000
			MIN	1,800	5,400
1858-1900	2,000[a]	400 to 800	MAX	2,903	3,450
			MIN	2,250	2,220
1900-1930	c.1,300	250 to 350	MAX	2,068	2,482
			MIN	not recorded	
1930-1943	c.1,300	100 to 250	MAX	not recorded	
			MIN	416	908
1943-1980	c.1,300	250 to 400	MAX	3,219	3,682
			MIN	1,416	1,139

Note: [a] 1,200 approved claims included many 'block' claims by large estates, including up to 150 tenancies.
Source: Countryside Commission Report 1984.

do more than speculate about the interactions between herbivores and vegetation in the past, but the *current* relationship between animals and vegetation can be more objectively assessed. The next four chapters will thus review what is known of the ecology and behaviour of the main herbivore species within the New Forest today, and their impact upon their environment. From such study and the understanding it gives us of the interrelationships between the herbivores and their vegetation, we can hope to design effective management for the future.

We must emphasise that what we shall attempt is a description of the current functioning of the New Forest ecosystem and the interrelationships of its various herbivores, both with each other and with the vegetation. In this chapter we have considered the history of the Forest and the influence of herbivores in the past. We now turn to a consideration of the ecology of the Forest in the present day, and it must be stressed that what we shall present in the succeeding pages is a description only of the *current* situation. From the studies to be described, restricted as they are to the ecology of the Forest as it is now, we can draw conclusions primarily about the present influence of the larger herbivores within the system; we may, however, from the understanding we gain of the underlying principles of the relationships between herbivores and vegetation, also make intelligent deductions about the effects of past grazing pressure (Chapters 7 and 8).

Chapter 3

The Grazers: Ecology and Behaviour of the Common Stock

The most abundant of the animals pastured under Common Rights today on the New Forest are the cattle and ponies. While the cattle are for the most part common everyday beef breeds, such as Hereford cross and other crossbreeds, turned out for the rough grazing the Forest affords, the New Forest pony has more of a tradition attached to it. Indeed, no-one knows precisely how long ponies have run upon the Forest. Certainly they are mentioned in Domesday Book — though it is not clear whether even by this stage these were domestic animals turned out to graze by private owners, much as they are today, or whether the Forest area may have supported truly wild stock — and it seems probable that the Forest has supported its own distinctive breed of pony for many centuries. Valerie Russell writes:

> In the days when the moors and forests of Southern England stretched practically unbroken from Southampton to Dartmoor, and even possibly to the fringes of Exmoor, wild ponies are believed to have wandered freely over the area. With the advance of civilisation sections of the Forest and moors were cultivated so that the ponies were restricted much more to the areas from which the present-day breeds take their name. (Russell 1976)

The New Forest Pony

The New Forest breed has suffered many attempts at 'improvement' over the centuries, so that it is now difficult to describe a distinctive type, but the typical Forest pony is a rather small, stocky animal, with an unfashionably large and coarse head; usually bay and somewhat shaggy in appearance, these are hardy, sturdy animals, shaped by the rigours of the environment in which they live as much

27

as by the breeding of those keen to 'improve' the stock. Market pressure has seen many such attempts to alter the characteristics of the breed over the years: the chief aim of the improvers was to introduce some quality into the New Forest ponies and particularly to increase their size. To this end, at various times Arab and thoroughbred stallions were used on Forest mares — and even stallions of hackney and carthorse blood. Almost without exception such attempts at improvement have led to a larger, but less hardy stock, ill-adapted to withstand the pressures of year-long subsistence on the Forest. Although such cross-breeding has left its mark on the Forest stock — and it is possible even now to identify animals of distinctly Arab type or Welsh type on the Forest — the effects were less marked than might be anticipated. In the 1930s came a ban on further infusions of outside blood, and such is the power still of *natural* selection upon the Forest that the New Forest pony has assimilated all these outside influences but still remains today a sturdy, cobby, shaggy bay not unlike the unimproved Exmoor (Plate 1). In recent years, there has again been increased selection for more 'sophisticated' stock; better-quality stallions are selected for release on the Forest — and are run only from May till September, so there is no need for the hardiness required if an animal is dependent upon the Forest all year. In result, the Forest pony once more is becoming more variable, more mixed in breeding, and it is again rather difficult to define a clear distinctive 'type'.

Most of the attempts to improve the Forest ponies have been prompted by market requirements. There is no doubt that the ponies grazed upon the Forest initially were the Commoner's own stock for use on their small holdings. As time progressed, however, the breed was developed for sale in various directions as outside markets came and went. Before the general use of wheeled transport, they found a ready market as pack animals; when wheeled vehicles became more widespread, the Commoners had another market for their ponies and Russell (1976) considers that it was chiefly as draught and harness animals that they were used until the early days of this century. With the coming of the Industrial Revolution, many of the native breeds were in great demand as pit ponies in the coal workings; the Forest ponies, however, were in general too large for such work and, while some of the smaller, stockier individuals went to the mines, the majority remained more popular as harness ponies. More recently, the demand for riding ponies, particularly for children, has opened up another lucrative market: the current attempts to upgrade Forest stock are directed chiefly at modifying the 'Forester' better to suit this new demand.

Allegations are still made from time to time that many of the Forest ponies are sold to the meat trade. Certainly, during the war years, many *were* bought for human consumption, and, when prices are low enough, large numbers are doubtless sold for the pet-food trade. Whether or not there is at present an export market to the Continent as has been suggested is, however, more doubtful. Russell notes:

> While it cannot be stated categorically that this has never happened, a certain amount of sleuthing has been done, and so far no proof has been forthcoming. In 1965 the council of the pony society asked for any evidence of such a trade, and the National Pony Society offered a substantial reward; but nothing even remotely conclusive was produced. With transport costs at their present level, it does not seem an economic proposition for foreign buyers to deal in New Forest ponies which, for the most part, do not carry much meat.

Social Organisation and Behaviour

Studies of the social organisation of most populations of wild or semi-wild horses (e.g. Feist and McCullough 1976; Berger 1977; Welsh 1975; Wells and von Goldschmidt–Rothschild 1979) have shown that the normal social 'unit' is a stallion-maintained harem of mares and sub-adults: a harem which is maintained throughout the year. Stallions are aggressively territorial and defend not only their group of mares, but also an exclusive home range (Feist and McCullough 1976).

The ponies of the New Forest, while equivalent in many respects to these feral populations, are more intensively managed. Groupings are often artificial, resulting as fixed associations of animals belonging to a single holding which were put out on the Forest together and retain their association. In addition, many owners do not leave their animals out all year but take them off the Forest over the winter, returning them to free range the following spring; such a practice reinforces each year the social bondings of the animals of a given holding. Not all owners, however, do take their stock in over the winter period, and many mares and foals — perhaps the majority — remain on the Forest throughout the year. None the less, 'natural' groupings are difficult to maintain: while many owners may be prepared to leave their mares and foals on the Forest all winter, few indeed allow stallions to overwinter. With a few exceptions, stallions are run on the Forest only from April to September. As a result, natural harem groups are rare. Females may form permanent associations (maintained throughout the year), and a stallion may join up with such a group over the summer months — but he plays little role in social cohesion: the groups are not true harems and the social order is essentially matriarchal.

The social structure of New Forest ponies is thus somewhat atypical, with the basic social unit most usually observed being a mare-foal assembly (Tubbs 1968, Tyler 1972). The fundamental elemental unit within such an assemblage may be recognised as an adult mare with her offspring of the current and previous years. Larger groups may be formed as associations of these basic units, but the size, cohesiveness and persistence of these social groups differ markedly. Stephanie Tyler, who published in 1972 the results of a classic study of the New Forest ponies and their social behaviour, recognised three main group types: (i) 'simple family units consisting of one adult mare alone or with one or more of her own offspring'; and (ii) groups of 'two adult mares with a varying number of offspring', or (iii) more rarely of 'three to six mares and their offspring'. The various proportions of these different types of grouping found by Tyler are shown in Table 3.1; the relative proportions of the different group types were not significantly different from those encountered by Pollock (1980) or in our own studies (Putman *et al.* 1981).

The most frequently observed social unit in New Forest ponies is Tyler's type (i): a family group consisting of a mature adult mare with one or possibly more of her past offspring and her current year's foal. Such a unit may vary in size from two, i.e. adult mare and

Table 3.1: Social Organisation of New Forest Ponies

Relative frequency of the different social groups of New Forest ponies recognised by Tyler (1972)

Winter	1 adult mare + offspring	2 adult mares + offspring	3 or more adult mares + offspring	Total no. of groups
1965/66	60.7	27.8	11.5	122
1966/67	59.0	28.2	12.9	124
1967/68	64.5	25.0	10.5	124

Due to the formation of false family groups (see p. 29), the true mother of some of the 'offspring' may be in doubt.

juvenile (plus perhaps a foal), to three or four where older offspring remain with the matriarch even as adults. This category blends almost imperceptibly into Tyler's type (ii) if older offspring, now themselves mature, remain with the group. Groups of more than three or four adult mares are, however, rare. This may be due in part to management intervention: many of the younger animals are sold each year, disrupting the social group and perhaps preventing the formation of larger groups of related adults. But this is not the complete answer. Tyler has shown that even under natural conditions most fillies do eventually leave their mothers' groups, usually at between two and four years of age. Such younger mares either join up with another group or with a single mare, or remain on their own to form the basis of a new family unit (Tyler 1972).

Not all larger groupings are of related animals; as noted earlier, association may be formed between unrelated animals belonging to the same owner. Such animals are generally kept in the same stable or paddock while off the Open Forest and relationships developed then will hold when the animals are released. Groups of two or three such individuals are common, and owners are widely aware of the extent of the bond between them and will rarely separate such groups deliberately. Such a phenomenon is also recorded by Arnold and Dudzinski (1978).

Nor are all associations as permanent as those described. Clearly, other more temporary associations are also observed. Many groups may co-occur on favoured feeding grounds, associating casually, merely because they happen to be in the same place at the same time. Further, groups of animals occupying the same home range frequently seem to use the range in the same way on a diurnal basis. This may give rise to what appear to be larger 'groups' of animals, but such aggregations are not in fact true social groups and do not persist for long. These aggregations of groups into larger units appear to be more common where larger home ranges are occupied.

Stallions, when on the Forest, may also form the focus of larger aggregations of mares, ranging in size up to 10-15 adults and juveniles. These harems, too, are simply composed of assemblages of mare groups and the group structure described above holds fast within the herd, i.e. members of a group will all form part of a single herd, and if one should leave then the others will also. Stallions' herds vary seasonally in size, being largest at the height of the breeding season, i.e. May to July, when the stallions actively herd as many mares as they are capable of holding together. Later in the summer, and particularly in winter, the stallions are less active and the groups comprising the herd appear to remain together on a more voluntary basis, individual groups coming and going from the herd as they please.

No significant differences in mean group size or in the distribution of Tyler's three group types is apparent within different geographic regions of the Forest. While size of social group does not vary over the Forest, group cohesiveness, however, does show marked variation. In a study of group size and cohesiveness in New Forest ponies all over the area, O'Bryan *et al.* (1980) showed that cohesiveness (defined in terms of permanence of association and median distance of each animal from others in its group) was markedly higher in a coarse-grained environment (areas where vegetational communities occur in large homogeneous blocks) than in finer-grained areas (where component vegetation types occur in a much smaller-scale mosaic). These areas were defined quite rigorously using Simpson's diversity index to delimit fine-grained (diversity index > 5.8) and coarse-grained ($2.8 < $ diversity index > 3.5) environments. Differences in group cohesiveness are highly significant: groups in coarse-grained environments remained in much closer association, both spatially and temporally, keeping fairly close together, and over longer periods of time than did ponies of fine-grained areas. Ponies in fine-grained areas operated far more independently, even within a group, and in many cases the true group was established only when the animals gathered together at dusk to move off into woodland areas for the night; median distance between group members was significantly higher in these areas. Further, it was noted that actual social interaction was reduced in fine-grained areas, social reinforcement in whinnying or allogrooming being far more evident in groups in coarse-grained environments. (Some possible reasons for the difference in group cohesiveness are presented by O'Bryan *et al.* 1980.)

Social and Sexual Behaviour

A hierarchical system, apparently based primarily on age and to a lesser extent on the character of the individual, operates both within social groups and between individuals of different groups (Tyler 1972, Gill 1980). Within a group, aggressive interactions are relatively uncommon and are used primarily only to reinforce the dominance hierarchy or if, for example, a third party takes too much interest in a newborn foal. Ear-threats and head-threats (a movement of the head towards the opponent, often with ears laid back) are the most common postures of aggression; if these displays prove ineffective, a mare may turn and threaten to kick her opponent with the hind legs (Gill 1980). Most disputes are resolved without need to press home an attack; but mares not infrequently bite, or kick out at each other in such encounters. Curiously, neither the degree nor the frequency of aggression shown by individual mares relates in any simple way to dominance position. Gill (1980) studied the frequency of the different aggressive encounters among 20 individual mares of one of the few permanent harem groups on the Forest, relating aggression to dominance hierarchy. There is little obvious relationship between dominance rank and frequency of aggression, although high-ranking animals are more inclined to press home a full-scale attack.

Friendly interactions within groups are much more common, with members of a given social 'unit' frequently nuzzling each other or grooming. Mutual grooming (allogrooming) is particularly common when two individuals have been feeding some distance apart for a period of time, or have been separated for some other reason, and seems to serve the social function of reinforcing the bond between the two, as well as the practical one of grooming those regions of the body which the individual cannot reach. (The importance of this last is suggested by the high incidence of allogrooming in the spring months at the time of the moult: Tyler 1972, Putman *et al.* 1981.) Few other direct interactions were observed between the members of a group, but such animals generally kept close company and would rest or lie up close together during periods of inactivity. If separated, the individuals show clear signs of distress, whinnying and searching actively for their associates.

Social facilitation was apparent in pony behaviour, although less so than in cattle, and applied both within and between groups. This partly explains the association of individuals which have overlapping home ranges, since a move by one individual will often precipitate several others to follow. Eliminatory behaviour appears to be influenced to some extent by social facilitation, although this more commonly within groups, and this is particularly true of eliminatory behaviour of stallions, who may 'cover' dung or urine of harem mares (Tyler 1972, Gill 1980).

Another important feature of group behaviour clearly influenced by facilitation is the curious phenomenon known locally as 'shading', and first described by Tyler (1972). Shading is a behaviour in which individual animals and even separate social groups come together and stand inactive in large congregations (commonly of 20-30 animals, occasionally numbering over 100) (Plate 5). It is particularly common in mid- to late summer, when shades may be formed as early as 9 a.m. and can last up to six hours. While shading, the animals stand close together, and remain largely inactive, merely whisking their tails or occasionally shifting position; no feeding occurs. When the shades break up, the animals drift away in their original social groups — but they may re-form the shade later the same day or on the following day; they re-form it on the same traditional site. The term 'shade' is in fact a misnomer, since animals do not necessarily seek shade from the sun (although many sites are in woodland, under rail bridges etc.) but may congregate on the brow of a hill to take advantage of any breeze which is blowing, or on such areas as roads or car parks which lack vegetation cover. The function of the behaviour is uncertain, but it is believed to be an adaptation to avoid the attacks of biting insects. (A similar behaviour in response to insects is described for Camargue horses by Duncan and Vigne 1979, and certainly the choice of 'shade' sites away from vegetation or in windy exposed locations would be compatible with such a function.)

A recent review of sexual behaviour in New Forest ponies has been presented by Gill (1980).

Home Range

The majority of pony groups on the Forest may be ascribed to a population whose focus of activity is a re-seeded or streamside lawn or other area of rich grazing. Since such lawns are relatively few in number, the populations are fairly discrete and few individuals or groups move regularly between populations. Each group of ponies occupies a well-defined home range — an area of the Forest to which it restricts its activity over the course of the day. Groups seldom venture outside this home range unless disturbed, although apparently exploratory forays were observed and mature mares may leave the area if no stallion is present when they come into oestrus. The home ranges of different groups *within* a population overlap extensively and are in many cases identical, and the pattern of habitat use of groups with overlapping ranges is often very similar. As noted above, this may lead a temporary aggregation of animals into large 'apparent' groups.

The size of the home range varies markedly over the Forest, and is usually determined by the proximity of four necessary components: a re-seeded area, streamside lawn or some other suitable grazing area;

a water supply; shelter, usually provided by woodland or gorse brake; and a shade. In the north of the Forest, where vegetation communities occur in relatively large, homogeneous units (coarse-grained), these components of the home range may be widely separated, and consequently large ranges often exceeding 1,000 ha may be found. Further south, a more heterogeneous vegetation cover results in smaller home ranges, down to 100 ha or less. No correlation has been observed between group size and home-range size, but the aggregation of groups into larger units appears to be more common in the north.

Pollock (1980) presents no data for range size of mares or mare groups, although he cites areas of 140 ha and 125 ha for two stallion groups. Range sizes of other free-ranging horses are reported in the literature (e.g. Klingel 1974; Feist and McCullough 1976). For two harem groups of free-ranging Exmoor ponies, Gates (1979) records range sizes of between 240 ha and 290 ha but notes that most activity was in fact restricted to a core area of only 45-60 ha.

Within this home area we can appreciate a general pattern of range use. Over the summer period, each animal generally spends most of its time feeding on the grazing area(s) included within its range. During the course of the day there is at least one move off the lawn to drink, and at this time the animals usually make some use of the lusher forage available in the wetter vegetation types. At dusk, there is a general move away from the more exposed grazing areas to vegetation types offering more cover, and the ponies usually spend the night in, or in the shelter of, areas of woodland or extensive gorse brake. Those groups of animals left on the Forest over the winter make somewhat less use of the exposed grassland communities, although these are still extensively grazed despite the fact that there can be little useful forage left. Over the winter, however, the ponies make greater use of those areas of their range offering shelter from the more severe weather, and emerge from the woodlands and gorse scrub for shorter periods during the hours of daylight. Range size is often reduced during the winter months, since water is more readily available and there is no need to 'shade'.

Such a description of range-use patterns is of course crude, over-general and over-simplistic. Detailed studies of the pattern of use of different vegetation types by the ponies were one of the main objectives of the programme of research sponsored by the Nature Conservancy Council between 1977 and 1981, and will be discussed at more length below.

Cattle

Cattle are now run on the Forest mainly for beef production, and are mostly Friesian/Hereford or more mixed crossbreeds. Individual herds of Red Devon and Highland also occur, however. Herds are composed predominantly either of one- and two-year-old heifers or mature beef cows and calves, these groups being commoned separately by their owners.

In contrast to the ponies, the basic social unit in cattle on the Open Forest is the herd, although the cohesiveness of this unit varies seasonally. The dispersion of herds over the Forest to some extent reflects the location of farmsteads, since many owners release their animals onto the common land during the daylight hours only and allow them to return to the enclosed fields at night. The majority, however, common their herds on the Open Forest throughout the summer months, generally releasing them on areas of good grazing such as re-seeded or streamside lawns. As in the case of pony 'populations' therefore, cattle herds belonging to different owners are dispersed over the Forest as relatively discrete units, each unit having a home range focused on one or perhaps more primary grazing areas.

Over the summer months, the herd fragments into a number of smaller groups which range independently but encounter one another frequently on grazing sites. These small sub-units are not cohesive and individuals regularly move from one to another on these encounters. The pattern of habitat use over this period is similar to that of the ponies, with cattle spending most of the daylight hours on the primary grazing areas, although making more frequent visits to water supplies. Less use is made of the wetland vegetation resources, and cattle appear to avoid particularly boggy areas. By night there is again a general move to those vegetation types offering cover, particularly deciduous woodland, although groups will often spend the night on open dry heathland when weather conditions are mild.

Cattle range more widely than ponies on the Open Forest over the summer-month period, often occupying home ranges with more than one primary grazing site. These areas may be used in turn over the course of several days, and daily movements of 4 km or 5 km are not uncommon. Movement is particularly common at dusk and dawn, groups often moving 2 km or more from the feeding site to the overnight area. These distances, although large in comparison to most pony-group movements, are less than have been recorded in similar studies on cattle in extensive ranges. Schmidt (1969) followed

Shorthorn cows over 9.2 km per day on open range, while Shepperd (1921) recorded daily movements of 8.9 km in a 260-ha paddock. This difference may be explained by the relative heterogeneity and availability of water in the Forest habitat. Cattle have been recorded as travelling 26 km daily in drought conditions on rangeland (Bonsma and Le Roux 1953).

Over the autumn months the majority of cattle are removed from the Forest in a series of 'drifts' or round-ups, and those few herds which are overwintered on the Open Forest are generally fed to a considerable extent with hay and straw. The practice of supplementary feeding radically alters the activity and ranging behaviour displayed by these animals. Sub-units of the herd are drawn together daily at the feeding site, and herd cohesiveness is markedly higher over the winter months. Feeding usually takes place on a fixed site one or two hours after dawn. The animals congregate at the site over this period, and stand around ruminating or inactive; little feeding takes place. With the arrival of the fodder the cattle will feed steadily until it has been consumed, this often taking as long as two hours or more, and will then lie ruminating over much of the remaining morning and early afternoon. Sporadic feeding may occur over this period, but total intake is negligible. During the mid to late afternoon the number of animals foraging or standing around increases, until there is a general departure from the feeding site around one hour before dusk. A steady movement then follows as the herd makes its way to the favoured overnight area. While the route taken may vary, the site selected to spend the night is almost always the same and this often involves considerably greater movement than occurs during the summer months. Round trips of over 11.5 km were recorded regularly. While en route, animals feed extensively, mostly on *Calluna* on dry and wet heath communities; they generally arrive at the night site as dusk falls. Feeding may continue over the first hour or two of darkness, particularly where a grazing area is available, before the animals bed down for the night. Despite the long hours of darkness, there is still very little feeding during the night. At first light the herd begins the return march, on which little foraging takes place, animals arriving at the feeding site in good time for the daily supply of fodder.

The reasons for this diurnal 'migration' are not immediately apparent. It may be assumed that the requirements of the overnight site are more rigorous during the winter months, and that relatively few suitable sites are available. Supplementary feeding generally takes place somewhere in the vicinity of the owner's farmstead in order to minimise the transport involved, and in selecting the site little consideration is given to the pattern of animal movement over the remainder of the 24 hours. Feeding sites and suitable overnight sites may therefore be widely separated, a daily 'route march' thus being enforced (Putman *et al.* 1981).

Patterns of Habitat Use

Within each animal's home range is contained a diversity of different vegetational types. The animal's use of its home-range area (pages 34-5 and 36-7) is clearly a consequence of the way in which it uses different vegetational communities which are available to it; the pattern of use of these various habitat types often shows considerable subtlety. In order to explore in more detail the way in which a cattle-beast or pony uses its vegetational environment and exploits the diversity of pattern, we must first examine more carefully the various different habitats available within the New Forest.

The Different Vegetational Communities of the Forest
In the first chapter of this book we presented a brief description of the Forest vegetation, emphasising that, despite its name, this is no forest of trees, but a hunting Forest of a wide diversity of habitats: some wooded, others open tracts of heathland or grassland. The diversity of habitat is indeed tremendous: a rigorous botanist would define any number of distinct formations, any number of separate systems with different characteristic associations of species. For our purposes, however, we must generalise to some degree (after all, reduced to its extremes no one square metre of the area is exactly the same as another), and in our studies of habitat use by the large herbivores of the Forest we have recognised 15 *major* habitat types. These may be grouped into gross categories as grasslands, heathlands and 'cover' communities (Table 3.2).

The natural *acid grasslands* of the New Forest, established for the most part on plateau gravels, comprise a total of about 3,000 ha of the Forest, mainly as small, irregularly shaped areas bordering deciduous woodland or heath and interspersed with clumps of gorse or thorn scrub. They are rather species-poor areas, generally dominated by the bristle bent (*Agrostis setacea*) and purple moor-grass (*Molinia caerulea*) and usually invaded to a greater or lesser extent by bracken (*Pteridium aquilinum*) and heather (*Calluna vulgaris*). On the more

Table 3.2: The main vegetation types recognised within the New Forest in this study, with details of soil, plant species composition, management history and extent in Forest as a whole (modified from Pratt et al., 1985)

Vegetation type (code)	Approx. total area in N.F. (ha)	Area of individual sites (ha)	Soil characteristics			Most abundant plant species	Remarks (History, Management, etc.)
			Parent material	Soil type	pH		
GRASSLAND							
a) Reseeded lawn (RL)	350	10-30	S,G	B	5.0-5.5	Agrostis capillaris, Festuca rubra, Bellis perennis, Hypochaeris radicata, Plantago lanceolata, Trifolium repens.	Established late 1940s on former A.G. Ploughed, fenced. Fertilised and sown with grass/clover mixtures. Fences removed 1-2 years after reseeding. (Browning 1951)
b) Commoners' improved grassland (CI)	270	3-10	S,G	B	5.2-5.7	A. capillaris, F. rubra, Poa compressa, Hieracium pilosella, sparse Pteridium aquilinum.	Established early 1970s on former A.G. Bracken swiped and limed.
c) Streamside lawn (SL)	320	< 10	A,C	B(Gl)	4.5-5.0	A. capillaris, Agrostis canina, Carex panicea, Juncus articulatus, H. radicata, T. repens.	Narrow lawns on flood-plains of partially canalised streams. Invasion of scrub (Prunus spinosa) common.

Key A = Alluvium, C = Clay, G = Gravel, H = Peat, S = Sand, B = Brown Earth, Gl = Gley, P = Podsol.

Table 3.2 continued

Vegetation type (code)	Approx. total area in N.F. (ha)	Area of individual sites (ha)	Soil characteristics Parent material	Soil type	pH	Most abundant plant species	Remarks (History, Management, etc.)
d) Roadside verge (RV)	170	2 m wide strip	G	—	6.0-7.0	*A. capillaris, F. rubra, P. compressa, Plantago coronopus, Trifolium spp., H. pilosella, Taraxacum officinale.*	Strips bordering unfenced roads; more extensive near car parks. Usually alongside heathlands with associated band of gorse (*Ulex europaeus*). Very shallow soils.
e) Acid grassland (AG)	3,400	< 10	S,C	P,B	3.9-4.6	*A. capillaris, F. rubra, Sieglingia decumbens, Agrostis setacea, Molinia caerulea, Carex spp., Potentilla erecta.* May be associated with *Pteridium, Calluna vulgaris* or *Ulex europaeus.*	Very variable. Gorse abundant on some sites. Bracken-dominated areas characteristic of better soils between dry heathland and deciduous woodland.
BOG	820	10-30	C,A	H	3.7-4.8	Eutrophic bogs: *Alnus* or *Salix* carr.	Variable depending on nutrient status of

Key A = Alluvium, C = Clay, G = Gravel, H = Peat, S = Sand, B = Brown Earth, G1 = Gley, P = Podsol.

					pH	Vegetation	Notes
						Oligotrophic bogs: *Sphagnum* spp., *Molinia, J. articulatus, Eriophorum angustifolium* etc.	catchment area. Often flooded and access difficult. Some sites modified by drainage, chiefly in 19C.
HEATHLAND a) Wet heath (WH)	2,020	5-20	C,A	P(Gl)	3.9-4.1	*Calluna, Erica tetralix, Molinia.* Scattered *Pinus sylvestris* saplings.	Often forms gradation from dry heath to bog or streamside lawn. Increasing proportion of *Molinia* and *E. tetralix* in wetter parts.
HEATHLAND b) Dry heath (DH)	7,380	10-50	G,S	P	3.7-4.0	*Calluna, Molinia.* Locally *Erica cinerea.* May be associated with *Pteridium, Ulex* etc., especially around margins.	Managed by burning on c.12-yr rotation. Recently burnt acres have high proportion of *Molinia* (regenerating heath-RH).
WOODLAND a) Deciduous woodland (DW)	7,940	<50	C,S	B(Gl)	4.0-4.5	*Fagus sylvatica* and *Puercus robur* with *Ilex aquifolium* understorey. *Betula pendula* and	Uneven age structure reflecting 3 main periods of regeneration associated with low

Key A = Alluvium, C = Clay, G = Gravel, H = Peat, S = Sand, B = Brown Earth, Gl = Gley, P = Podsol.

Table 3.2 continued

Vegetation type (code)	Approx. total area in N.F. (ha)	Area of individual sites (ha)	Soil characteristics Parent material	Soil type	pH	Most abundant plant species	Remarks (History, Management, etc.)
						Crataegus monogyna around borders with A.G.	grazing pressure. (1) 1650-1750, (2) 1850-1920, (3) 1935-1945 (Tubbs 1968). (See page 153)
b) Coniferous woodland (CW)	4,840	< 20	G,S	P(G1)	3.8-4.1	*P. sylvestris* with under-storey of *Ilex* and *Crataegus*. Sparse ground flora of *Pteridium* and *Calluna*.	*P. sylvestris* introduced in late 18C. Largely self-sown and spreading on heathland and woodland margins. Partially controlled by cutting.
c) Gorse brake (GB)	1,040	< 50	S,G,C,	B	4.0-4.5	*Ulex* with ground flora of *A. setacea*, *Molinia*, *Calluna* and A.G. species.	Narrow belts near lawns, roads, trackways, etc.; more extensive tracts associated with some A.G.s. Indicator of past human activity (Tubbs & Jones 1964). Partially managed by cutting or burning.

Key A = Alluvium, C = Clay, G = Gravel, H = Peat, S = Sand, B = Brown Earth, G1 = Gley, P = Podsol.

fertile alluvial soils at the margins of the many small rivers which traverse the Forest, these acid grasslands are replaced by the more lush grazing of *streamside lawns* (Plate 6). The flora of these grass-lands is variable, dependent on the degree of waterlogging/frequency of flooding etc. The dryer areas are dominated by grasses such as *A. tenuis, Cynosurus cristatus, Lolium perenne* and *Anthoxanthum odoratum*, whilst the damper parts have a greater abundance of *A. canina, Alopecurus geniculatus*, rushes (*Juncus bulbosus* and *F. articulatus*) and sedges (*Carex panicea* and *C. flacca*). Moisture-loving species such as creeping buttercup (*Ranunculus repens*), lesser spearwort (*R. flammula*), marsh ragwort (*Senecio aquaticus*), and marsh pennywort (*Hydrocotyle vulgaris*) are common, with plants such as water mint (*Mentha aquatica*), water-pepper (*Polygonum hydropiper*) and floating sweet-grass (*Glyceria fluitans*) abundant in drainage ditches and oxbows.

As noted earlier, the Forest also offers to the grazing herbivores a variety of artificial grasslands. A number of different management techniques have been attempted over the years, resulting in the formation of three distinct types of 'improved' grasslands: Common-ers' improved grasslands, Verderers' improved areas, and re-seeded lawns (Plate 7).

About 20 areas of acid grassland, gorse and heathland totalling 350 ha were ploughed, fenced, cropped for several years and re-seeded with grass during the period 1941-59 under the New Forest Pastoral Development Scheme. This increased the agricultural productivity of the Forest during wartime and provided improved grazing for the rapidly increasing population of commonable animals when the fences were removed. The seed mixtures and fertilisers used in creating these *re-seeded lawns* varied considerably in different areas, with rye-grass (*Lolium perenne*), cock's-foot (*Dactylis glomerata*) and Timothy (*Phleum pratense*) particularly common, accompanied by heavy applications of lime, chalk and super-phosphate. Other grasses frequently used included red fescue (*Festuca rubra*) and crested dog's-tail (*Cynosurus cristatus*).

Since re-seeding, the lawns have been colonised by indigenous grasses, e.g. *Agrostis tenuis, F. rubra*, and a wide variety of other grassland species, such as selfheal (*Prunella vulgaris*) and procumb-ent pearlwort (*Sagina procumbens*) which are abundant on the dryer soils, with buttercup (*Ranunculus repens*) etc. common in the damper areas. There is also an abundance of leguminous species, e.g. *Trifolium repens* (white clover), *T. dubium* and *T. micranthum* (yellow trefoils) and *Lotus corniculatus* (bird's-foot-trefoil). Many lawns show some signs of recolonisation by bracken (*Pteridium aquilinum*), heather (*Calluna vulgaris*) and gorse (*Ulex europaeus*), although these species are not usually vigorous.

Commoners' improved areas (a total of about 260 ha within the Forest as a whole, mainly in the south) were derived originally from native acid grassland. The grasslands were swiped to cut back the bracken, and then treated with lime, phosphate and other fertilisers as part of a limited scheme of pasture improvement carried out between 1969 and 1971. The flora of these areas reflects that of the acid grasslands from which they are derived, but with a greater species diversity as a result of improved soil conditions and reduced bracken cover. Several grasses are abundant, including *A. tenuis*, *Festuca rubra*, *Sieglingia decumbens* and *Poa annua*, and numerous herbaceous species are found, in addition to those also common in acid grassland, e.g. yarrow (*Achillea millefolium*), plantains (*Plantago lanceolata* and *P. major*) and dog-violet (*Viola canina*). Composites, such as the daisy (*Bellis perennis*), autumn hawkbit (*Leontodon autumnalis*), ragwort (*Senecio jacobaea*), cat's-ear (*Hypochoeris radicata*) and thistle, are common, and heathland species (*Calluna*, *Erica* spp., *Ulex* spp., etc.) are also found. A second series of improved grasslands (Verderers' improved areas) was established in the early 1960s, in the form of firebreaks totalling about 100 ha around ten Forestry Commission plantations. Although fenced, these strips are generally accessible to commonable animals via gates, and can be used to hold animals after rounding-up etc., unlike the main plantation firebreaks from which cattle and ponies are excluded.

Rotovating and re-seeding of these areas was less successful than the re-seeded lawn improvements, and many now support only a limited grassland cover. Drainage is often poor, with the damper areas dominated by sedges (*Carex* spp.), *Ranunculus* spp., composites (e.g. *Leontodon autumnalis*), and streamside-lawn species.

Although not deliberately 'improved', the Forest contains one other important type of artificial grassland in the many *roadside verges*, a grassland type quite widespread and extensively used by commonable animals. The verge normally occupies the 1.5-m-wide strip of disturbed ground between the road and the ditch (c. 30 cm deep) excavated to restrict the access of cars to the Open Forest. In areas of high human and animal pressure, e.g. around car parks and camp sites, the verge may, however, extend to about 50 m from the road before merging gradually with the surrounding vegetation.

The flora of the roadside verges is generally similar to that of the shorter areas of the re-seeded lawns, although more sparse and with a higher proportion of *F. rubra* and *S. decumbens* grasses. Composites and rosette plants such as yarrow (*A. millefolium*) and buck's-horn plantain (*P. coronopus*) are particularly common.

The transitional zone between the verge and the neighbouring vegetation (frequently heathland) comprises mainly composites (*H. radicata* and *Leontodon* spp.) with *Molinia* sp. and *Sieglingia* sp. grasses. The dwarf gorse (*Ulex minor*) is at its most abundant in this zone, whilst clumps of gorse and bramble are also typically found.

The other main vegetation types of the Open Forest are the various *heathlands* and the *valley bogs*. Vegetation dominated by heather (*Calluna vulgaris*) extends over about 4,500 ha of the Open Forest, mainly on the nutrient-poor podsolised sands and gravels (Plate 8). Growing in association with the heather are *Molinia* sp., dwarf gorse (*Ulex minor*), heaths (*Erica cinerea* and *E. tetralix*) and bristle bent (*Agrostis setacea*), these species being more abundant in the younger-aged communities. Mature dry heath is often invaded by bracken, particularly in the north of the Forest, whilst gorse bushes (*U. europaeus*) are common on disturbed soils, e.g. near roads, tracks, railway cuttings, etc.

The vegetation of these *dry heaths* is maintained as a 'fire-climax' by periodic (every 6-15 years) controlled burning. Alternatively, the mature heather may be removed by mechanical cutting and baling for use in road construction etc. These operations reduce the risk of accidental fires, and fulfil the statutory requirement of the Forestery Commission to prevent scrub invasion of the grazing lands, whilst supposedly increasing the amount of forage available for stock. The structure of the heathland clearly varies enormously with the time interval since it was last burnt or cut. Accordingly, we distinguished a series of different dry-heath communities within the Forest: recently *regenerated heath* and dry heath of ages 3, 7 and 12 years.

Humid and *wet heathland*, in which *E. tetralix* is abundant and often replaces *Calluna* sp. as the dominant dwarf shrub, occupies a further 2,000 ha (approx.) of the Forest. It frequently occurs as a transition zone between dry heath and bog or streamside lawn. *Molinia* is generally more abundant than in the dry-heath community, and other moisture-loving plants are also found, e.g. *Juncus* spp., *Carex* spp., and sundew (*Drosera* spp.). Such communities grade, in wetter areas, into the *valley bogs* so characteristic of the New Forest system (Plate 9). These bogs occupy about 900 ha of the

New Forest, the most extensive tracts occurring in the southern part of the area. The bog vegetation is very variable, depending on the acidity of the water flowing through it. Thus, water derived from the relatively alkaline Headon Beds and Barton Clays gives a central alder-carr vegetation (although the edges may be acidic), whilst soil water from the acid gravels and sands supports vegetation dominated by *Sphagnum* mosses and *Molinia caerulea,* with some local development of grey willow carr (*Salix cinerea*).

Cover communities are recognised within the Forest as woodlands and gorse brakes. Unenclosed woodland, of which more than 80% is deciduous, extends over about 3,350 ha of the New Forest, whilst the Statutory Inclosures, with 30-40% hardwoods, occupy a further 8,300 ha. The *deciduous woods* are dominated by mature oak (*Quercus robur*) and beech (*Fagus sylvatica*), usually with an understorey of holly (*Ilex aquifolium*), although hawthorn (*Crataegus monogyna*) occurs in a few places. The ground flora consists of glades of *A. tenuis* grass (recognised as a distinct vegetation type '*woodland glade*'), with species such as bracken, foxglove (*Digitalis purpurea*), bramble (*Rubus fruticosus*), wood spurge (*Euphorbia amygdaloides*) and wood-sorrel (*Oxalis acetosella*) locally common. The more densely shaded areas (particularly under beech), however support very little herbage. The unenclosed coniferous woods are dominated by Scots' pine (*Pinus sylvestris*) with a scant understorey of holly and hawthorn bushes etc. The ground flora is composed of bracken, bramble, honeysuckle (*Lonicera periclymenum*), ivy (*Hedera helix*) etc., with very little grass except on paths and rides. Finally, shelter is also offered to the grazing animals of the Forest in the extensive *gorse brakes* of *Ulex europaeus* which develop on acid-grassland or heathland communities or around the edges of the various improved grasslands.

Patterns of Habitat Use

Animals must seek from their environment satisfaction of a number of different requirements — food, shelter, water, access to mates — and the patterns of habitat use we observe reflect their attempts to satisfy these various needs from the limited range of communities and habitats available to them. The needs change, and the resources offered by different habitats themselves change; the observed pattern of use of habitat tracks and reflects these changes in resource quality and animal requirements. The Common animals of the New Forest have available to them a complex mosaic of vegetational communities as described above. Their use of this mosaic is best understood if we appreciate that their primary requirements from the Forest environment are food (and to a lesser extent water) and shelter. Differences in emphasis of these separate priorities at different times (a lesser need, perhaps, to worry about daytime shelter during summer than in winter) or changes in availability of the required commodity in different communities (as, for example, relative availability of forage may alter seasonally between the different habitats of the home range) will result in changes — diurnal or seasonal — in the observed use of habitat.

The resultant distribution of cattle and ponies across the various communities of the Forest are summarised in Figures 3.1 and 3.2.

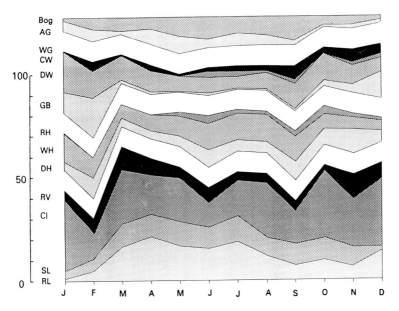

Figure 3.1: Seasonal changes in the distribution of ponies across the different vegetation types of the New Forest. The vegetation types have been grouped as follows: Grasslands: *re-seeded lawns (RL), streamside lawns (SL), Commoners' improved grassland (CI), roadside verges (RV);* Heathland: *dry heathland (DH), wet heathland (WH), regenerating heathland (RH);* Cover communities: *gorse brake (GB), deciduous woodland (DW), coniferous woodland (CW), woodland glades (WG);* Other vegetation types: *acid grassland (AG), bog. Data are expressed as percentages of total observations in each month. The scale on the left shows divisions of 10% total observations*

Source: Pratt et al. (1985).

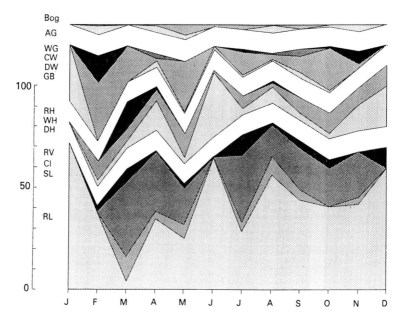

Figure 3.2: Seasonal changes in the distribution of cattle across the different vegetation types of the New Forest. See caption to Figure 3.1 for key to vegetation types

Source: Pratt et al. (1985).

Data presented here are drawn from detailed studies of the patterns of habitat use by cattle and ponies in the New Forest published by Putman *et al.* (1981, 1982) and Pratt *et al.* (1985). The data themselves derive in the main from two observational techniques.

1. *Static watches.* Two replicate sites were chosen for each of the vegetational communities defined above. In each month, one replicate of each community was observed over a 24-hour period (alternate replicates being observed in the following months). The observer recorded number and activity of each species present on the site at fixed intervals throughout the 24-hour period.

2. *Transect watches.* Transect routes, selected to run through as many different vegetation types as possible within a small area, were established in two main parts of the Forest. Set points, commanding a clear view of a particular vegetation type, were fixed for each route and observations were made at these predetermined points every two hours during the 24 hours of a day. Each transect route was watched

for a full 24-hour period every fortnight. (Methods throughout are described in detail elsewhere: Putman *et al.* 1981).

Results from all observational techniques confirm that, for the *ponies*, the various different grasslands of the Forest (particularly re-seeded lawns, streamside lawns and Commoners' improved grasslands) are tremendously important communities throughout the year (Table 3.3). Proportion of animals on improved grasslands and streamside lawns (as a proportion of total numbers observed on all communities) stays for the most part in excess of 40-50%. Increasing stock numbers, often coupled with the effects of drought, result in reduced availability of grass over the summer months. Declining use of the grasslands is compensated for by increased use of the bogs and both wet and regenerating heath; in autumn and winter, increased use is made of deciduous woodland and gorse brake. To a large extent, pattern of habitat use for feeding matches pattern of total occupance, suggesting that habitat use may be influenced primarily by foraging. During winter periods, however, and in hours of dark-ness at all times of year, extensive use is made of deciduous woodland edge and/or gorse brake for shelter. Feeding continues, but at this time choice of forage is dictated mainly by what is available within that community rather than by the converse.

Cattle were observed in all community types used by the ponies, although on all methods of observation far fewer animals were recorded (most are removed from the Forest during winter months). The pattern of habitat use revealed by transect watches show, as for the ponies, the importance of grassland communities. The various improved grasslands, together with streamsides, sustained in excess of 40% of cattle records for most of the year, with over 70% use in the summer months (July-September). Heathland communities were also used extensively throughout the year, except in midwinter months. Deciduous woodland showed peak use in spring and autumn; re-generating heath and bogs showed peak occupancy from March to May. Overall, the patterns of community use throughout the year shows relatively little variation, and displays considerably less flexi-bility, than that of the ponies (see Table 3.4).

Over the autumn months the majority of cattle are removed from the Forest in a series of 'drifts' or round-ups, and those few herds which are overwintered on the Open Forest are generally fed to a considerable extent with hay and straw. The practice of supplement-ary feeding radically alters the activity and ranging behaviour

Table 3.3: Use of habitat by New Forest ponies

	Jan	Feb	Mar	Apr	May	June	July	Aug	Sept	Oct	Nov	Dec		
	Percentage of ponies seen in each vegetation type each month													
RL	2	5	17	21	14	17	19	12	8	10	11	16	RL	(Re-seeded Lawn)
SL	5	6	12	9	19	12	13	9	9	11	13	3	SL	(Streamside Lawn)
CI	33	15	27	22	14	11	18	27	18	34	23	27	CI	(Commoners Improved Grassland)
RV	4	8	11	6	6	6	4	5	5	2	9	6	RV	(Roadside Verge)
AG	3	9	1	10	3	4	6	4	3	2	3	2	AG	(Acid Grassland)
DH	4	10	4	3	14	11	7	7	12	11	18	11	DH	(Dry Heath)
WH	15	9	7	7	14	14	13	13	13	7	5	8	WH	(Wet Heath)
Bog	2	5	7	9	9	12	8	10	12	4	4	4	Bog	
DW	20	12	12	10	6	6	6	9	13	13	7	6	DW	(Deciduous Woodland)
CW	0	0	0	0	1	3	2	1	2	0	0	2	CW	(Coniferous Woodland)
WG	0	4	0	2	0	2	3	2	6	2	4	5	WG	(Woodland Glades)
GB	12	16	2	1	0	1	0	0	0	3	3	11	GB	(Gorse Brake)

Table 3.4: Use of habitat by New Forest cattle

	Jan	Feb	Mar	Apr	May	June	July	Aug	Sept	Oct	Nov	Dec		
													RL	(Re-seeded Lawn)
	51	29	19	18	24	73	31	55	43	42	30	52	RL	(Re-seeded Lawn)
	24	0	8	6	16	0	5	9	5	0	2	11	SL	(Streamside Lawn)
	2	0	40	17	19	0	37	16	21	16	34	0	CI	(Commoners Improved Grassland)
	0	3	4	1	3	0	11	1	3	4	0	4	RV	(Roadside Verge)
	9	4	0	2	2	1	0	4	0	1	3	4	AG	(Acid Grassland)
	0	0	12	5	3	24	2	9	3	3	17	11	DH	(Dry Heath)
	0	5	5	16	9	1	6	2	7	10	7	6	WH	(Wet Heath)
	0	1	0	4	5	0	2	0	2	0	0	11	Bog	
	0	39	10	10	16	0	9	2	8	24	2	0	DW	(Deciduous Woodland)
	1	0	1	0	0	0	0	1	6	0	0	0	CW	(Coniferous Woodland)
	0	18	0	1	0	1	1	0	0	1	5	0	WG	(Woodland Glades)
	13	0	0	1	2	1	1	0	2	0	0	0	GB	(Gorse Brake)

Percentage of cattle seen in each vegetation type each month displayed by these animals. Sub-units of the herd are drawn together at the feeding site daily, and herd cohesiveness is markedly higher over the winter months (see page 37).

Both herbivore species display a pronounced diurnal shift in habitat choice within the overall pattern of community use described, with daytime habitats selected primarily for feeding and night-time selection for communities offering some degree of cover. This shift appears to be correlated with the daily patterning of activity in the two species: foraging is undoubtedly the major daylight activity for both cattle and ponies, with the time devoted to resting (or rumination) increasing after dark (Pratt *et al.* 1985).

Diurnal movements between vegetation types by *ponies* are very much more marked over the summer months than in winter. Throughout the year, ponies show preferential night-time use of vegetation which provides cover or shelter. In winter, however, these 'shelter' communities are also used more extensively during the daylight than they are during the summer. Thus, over the winter months, the ponies spend much of the time in sheltered areas, making periodic forays onto the more open vegetation communities, and radical day/night changes in dispersion are uncommon. In summer, however, most of the daylight hours are spent in open habitats (grasslands, heathlands and bog communities), and diurnal movements into sheltered areas are very much more obvious.

Day/night differences in cattle dispersion are even more marked than those observed in ponies and persist throughout the year. In all seasons, cattle spent most of the daylight hours on the open vegetation types, and particularly the grasslands, but made a deliberate move onto other communities at dusk and back at dawn. Vegetation types most commonly used by night were deciduous woodland and dry heath.

Clearly, animals are not evenly distributed across the various communities available to them within the Forest. This, together with the obvious movement at dusk and dawn to and from communities offering shelter, makes it clear that animals positively select which habitats to occupy in their home range at a given time. Animal *preferences* for habitats may be considered in terms of a preference index calculated as the degree to which dispersion between different community types differs from a purely regular distribution (with numbers in each habitat merely reflecting relative areas of that habitat in the environment as a whole). Thus, an index

$$\frac{\text{Number of animals observed in community A}}{\text{Area of A}} \times \frac{\text{Total area surveyed}}{\text{Total number of observations in } all \text{ vegetation types}}$$

will equal 1 *if* animals are evenly spread across all vegetation types, will be > 1 if positive preference is shown for a particular community, and < 1 if the community is avoided. In Table 3.5, this index is presented as a logarithm; this adjusts the values to a more easily visualised scale so that a score of zero now represents no preference, while any positive number is positive preference and any actual avoidance will appear as a negative value in the tables. These data confirm the seasonal trends in pattern of use described above, but reveal preferences for individual community types which are not necessarily apparent from the pure analysis of animal distribution (Figures 3.1, 3.2). Among ponies there is a strong positive preference for grasslands throughout the year ($P_i = 0.410$ February to $P_i = 0.755$ March), but the index shows that streamside lawns and roadside verges are in fact preferred to Commoners' improved and re-seeded lawns despite the fact that these latter grassland types support more animals. In contrast, all heathland communities are under-exploited (greatest preference was in June and September, when P_i was -0.420). No strong preference for or against the woodland and gorse-brake communities is apparent over the summer months, but they are increasingly favoured during the winter when shelter is more important (max. 0.505 February). Conversely, the lowland bogs are avoided over the winter months but show positive preference in summer. Preference indices calculated for *cattle* confirm a more confined use of the habitat than occurs in ponies and with even greater preference being shown for the improved grasslands. In this case, however, preference for individual grassland types corresponds to the observed distribution of actual observations, with re-seeded lawns and Commoners' improved areas favoured over streamside lawns and roadside verges. Positive preference is also shown for the cover communities, primarily deciduous woodland, in several months of the year. As with ponies, the heathland communities are relatively under-exploited.

Habitat preferences and thus individual patterns of habitat use differ between individual animals. What we have described here is an overview of the pattern of use of vegetation by a whole population;

Table 3.5: Preferences shown for different vegetation types by New Forest Ponies and Cattle

PONIES

Vegetation	J	F	M	A	M	J	J	A	S	O	N	D
GRASSLANDS	0.544	0.410	0.755	0.716	0.694	0.629	0.689	0.654	0.522	0.633	0.590	0.635
HEATHLANDS	−0.538	−0.495	−0.770	−0.745	−0.585	−0.420	−0.509	−0.495	−0.420	−0.538	−0.481	−0.744
WOODLAND and GORSE	0.394	0.505	0.107	0.134	−0.108	0.127	0.041	0.053	0.290	0.199	0.228	0.336
BOG	−0.721	−0.194	−0.143	−0.187	0.025	0.053	−0.060	0.041	0.143	0.292	−0.252	−0.602
ACID GRASSLAND	0.057	0.155	−0.469	0.250	0.299	−0.027	0.146	0.004	−0.237	−0.409	−0.187	−0.469

CATTLE

Vegetation	J	F	M	A	M	J	J	A	S	O	N	D
GRASSLANDS		0.524	0.702	0.777	0.653	0.774	0.839	0.843	0.792	0.710	0.712	
HEATHLANDS		−0.721	−0.444	−0.481	−0.620	−0.292	−0.921	−0.770	−0.770	−0.699	−0.469	
WOODLAND and GORSE		0.566	0.193	−0.268	0.377	−1.000	0.111	−0.523	0.100	0.286	−0.187	
BOG		−0.921		−1.222	−0.276	—	−0.602	—	−0.602	—	—	
ACID GRASSLAND		−0.013		−0.201	−0.161	−0.602	—	−0.036	—	−0.745	−0.114	

within this there is considerable individual variation (Howard 1979, Gill 1984). In part, such differences are imposed upon the animals because of differences in the relative availability of the various vegetational communities in different geographic regions of the Forest. Thus, pony populations in areas without extensive gorse brake will make more use of woodland for winter shelter, and ponies in areas with no improved grasslands will make more use of stream-sides or natural acid grassland and heath. But this is not a complete explanation: even among individuals who share the same basic range there are subtle differences in the extent of use of different habitats; such individual differences clearly reflect true differences in habitat preferences, and may have considerable implications in relation to differences in the ability of individual animals to maintain body condition (Gill 1984).

In practice, however, such differences in habitat use are rather slight; indeed, the overall impression is of surprising uniformity in community use. Studying patterns of habitat use by Forest ponies in four different sites chosen specifically because they *did* differ markedly in vegetation and standing crop, Gill found that habitat use by the ponies in the four areas was fundamentally rather similar. Ponies in areas which showed marked differences in relative availability — and juxtaposition — of different habitats, none the less end up displaying very similar patterns of habitat use.

Comparison of Patterns of Habitat Use of Cattle and Ponies

When the patterns of use of different vegetational types by ponies and by cattle over the course of the year are compared, a number of clear differences may be noted.

We may remark once more on the relative constancy over the year of the pattern of use of the different communities by cattle, which contrasts markedly with the flexible pattern of community use displayed by the ponies. Thus, cattle spend 60-70% of their time on grassland communities and 10-20% of their time in woodland (feeding on grasslands and heathland during the day and moving into woodland for shelter at night), and this pattern changes little over the year. To some extent this constancy of habitat occupancy — together with the choice of the actual communities themselves — may be a result of the strongly herding pattern of social organisation of the cattle: restricting activity to relatively few communities and making major changes in patterns of habitat occupancy hard to bring about. (Pony social grouping, like habitat use, is far more flexible.) It may also reflect the fact that cattle on the Forest are, in general, subject to a far greater degree of management by their owners than are the ponies. Patterns of habitat use may thus be a result, in large part, of owners' management policy.

Not only does the pattern of use of communities by the cattle show relatively little seasonal variability, but the actual range of community types occupied is also more restricted than that utilised by the ponies. Certain communities which are exploited extensively by the ponies are used but little or not at all by the cattle. While the ponies, during the course of a year, show some degree of occupancy of a wide range of community types, cattle restrict most of their time to 'improved' grasslands, heathland and woodland (primarily deciduous). Acid grassland, bog and gorse brake are used hardly at all by comparison, while regenerating heath and woodland glade (at certain times of year extremely important to the ponies) are virtually excluded.

Even where certain communities are used in common, they are not necessarily of equal importance to the two species. Cattle show a far greater use of grasslands (particularly areas of re-seeded lawn) in winter than do the ponies and, while ponies make more extensive use of wet heath throughout the year than do cattle, their use of dry heath is much lower. Further, even *within* grasslands, the two species show distinct preferences. Analyses of preferences above (Table 3.5) show marked preference by ponies for streamside lawns and roadside verges among the grasslands and also woodland glades. Cattle also favour grasslands, but are found predominantly on re-seeded and Commoners' improved areas. A partial explanation for this comes from the observation that cattle appear unwilling to occupy any grassland area which is less than 10 ha in size. Strongly social animals that move around their home range as a distinct herd, it may well be that areas less than 10 ha in area are insufficient to support the full herd. Such a constraint restricts them from use of road verges and streamsides, which are characteristically of a small available area. This difference in social behaviour between the strongly herding cattle and the essentially individualistic ponies may also underlie the observed differences in the pattern of range use of the two species. Ponies act very much as individuals and drift around their home ranges as they feed. The cattle by contrast act as an integrated herd; in addition the herd does not merely move as it feeds but makes quite distinct 'route marches' between different areas, moving from one feeding area to another — or to the cover of woodlands at night. These route marches are markedly purposeful and the animals may cover considerable distances (2-4 km). Such moves are quite distinct from the random drifting of movements during grazing. The animals have an almost inflexible circuit which they tour daily, as a herd, from shelter areas of overnight rest to daytime feeding sites, and back. The daily circuit is fixed and repeatable, encompasses the same areas and same vegetation types, and is followed faithfully. (These same factors also perhaps help to explain the clear differences between the two species in terms of seasonal variation in habitat use, already described. The pattern of habitat use by cattle changes but little, entrained as it is on the daily circuit; habitat use by ponies is, as we have seen, far more flexible, and shows great seasonal change in communities selected.)

Despite these various differences in the pattern of habitat use shown by cattle and ponies, the overriding impression is actually one of striking similarity. Both species are preferential grazers, feeding primarily on the various improved grasslands of the Forest and, over-night, seeking cover in woodlands or gorse brake. Even in winter, when the ponies are making more use of other foodstuffs (Chapter 4), 40-50% of observations are still recorded from areas of improved grassland; in the summer, both species again spend the majority of their time on grasslands (particularly RL, SL, CI) and wet heath. The various differences between the species to which we have just drawn attention are actually relatively *minor* differences in an overall picture of tremendous similarity. Calculations of overlap in the pat-terns of habitat use shown suggest an overlap of somewhere in the region of 78% (see Chapter 4). To what extent, then, may we consider cattle and ponies in the New Forest in competition?

From our considerations to date, it is clear that the major overlap in habitat use between cattle and ponies throughout the year is in the use of the various improved grasslands of the Forest. Both species rely heavily on these grazing areas — which are in fact a relatively limited resource within the Forest as a whole. In excess of 40% of all observations for either species are from these various grassland areas — in whatever month — yet these improved grasslands account for only some 1,140 ha in total area. We have already noted some degree of separation in the type of improved grassland favoured. Perhaps for social reasons (page 56), cattle make far less use of the Commoners' and Verderers' improved areas (and the natural streamside lawns) than do ponies and concentrate primarily upon re-seeded areas. But ponies also favour re-seeded lawns, and many such lawns support large popu-lations of both species simultaneously. Closer examination reveals, however, that despite this overlap — or perhaps because of it — ponies and cattle *are* separated even in their use of a single community.

The most detailed information we have on this spatial segregation within communities is for the Forest grasslands, and particularly for re-seeded lawns. In these areas, it is clear that cattle and ponies feed in and occupy distinct areas on the lawns; and that these areas differ botanically, both in structure and species composition (for fuller details see Chapter 8). Each lawn thus comprises a mosaic of distinct patches — which we term 'short grass' ($<$ 20 mm) and 'long grass' (20-50 mm) — occupied by ponies and cattle respectively.

The patches derive from the selective eliminatory behaviour of, particularly, ponies. Most large herbivores avoid feeding in the close vicinity of their own fresh dung, an adaptation presumed to reduce the risk of parasitic reinfestation: if an animal returns to graze an area on which dung has been deposited only after a considerable period (during which the dung has largely decayed), the risk of increasing its permanent parasitic burden to unacceptable levels is considerably reduced. Horses are particularly 'hygienic' in this regard and recognise specific latrine areas within their range, to which they move to defaecate. Such behaviour is a well-known feature of horses in pasture fields (Archer 1973, 1978; Odberg and Francis-Smith 1976, 1977), but it was originally thought that it occurred in response to enclosure and would not be shown by free-ranging horses in areas such as the New Forest. Indeed, Tyler (1972) suggested that there *was* no such pattern of selective dunging and grazing by the Forest ponies. More recently, however, Edwards and Hollis (1982) have clearly demonstrated that a mosaic of 'latrine' and 'non-latrine' areas does develop on New Forest lawns in response to the same pattern of selective feeding and dunging in the free-ranging ponies of our studies.

When feeding on lawns, many animals move off the area into adjacent vegetation types to defaecate. This is not, however, always possible, and other latrine areas are defined within the grassland community itself. Since the animals will not feed in these latrine patches, the grass grows longer in these areas (aided by the fact that the latrines obviously become considerably better fertilised than the rest of the sward). Thus a pattern is established, of ponies feeding on 'short-grass' patches (which are cropped as close as their teeth will allow) and defecating in 'long grass'. Since the mechanics of the way in which cattle obtain their food prevent them from feeding on very short grass, cattle on these same communities are forced to restrict their feeding activity to the 'long-grass' patches on the lawn. Cattle, unlike ponies, defaecate at random, so to speak; since they spend almost all their time in 'long-grass' areas, however, their faeces remain in these patches, and so the mosaic is maintained. As a result, use of grasslands is divided, with cattle occupying and feeding within 'long-grass' areas and ponies feeding on 'short-grass' patches.

This account is of course over-simplified to a degree. Over the winter, when animal numbers on the Forest are reduced (so that faecal accumulation in any area is also reduced), and when available grazing is also at a minimum, ponies do begin to graze into the long-grass areas. By the end of the winter, the mosaic pattern is less apparent and ponies feed and defaecate freely all over the lawns; indeed, at the beginning of the growing season they feed preferentially in the more nutritious long-grass areas. The pattern begins to re-establish itself, however, from April/May, and by June/July complete separation of cattle and pony feeding areas is achieved (Edwards and Hollis 1982). Although this pattern of spatial segregation is well established only on lawn communities, we have reason to believe that it may also occur, to a lesser degree, in most other communities (Putman *et al.* 1981). Its influence is in any case most important in the grassland types for which it is described here, since it is here that the greatest overlap in feeding use of communities between cattle and ponies is experienced.

In summary, pattern of use of the various communities differs markedly between cattle and ponies, in terms of the range of communities occupied and seasonal variation in community use. Both species are preferential grazers, however, and during the summer patterns of habitat use become very similar, with animals feeding in grassland and wet heath and moving into woodland at night. Some separation in feeding habitat at least is still maintained at this time, owing to increased use of wet heath by ponies and to the establishment of 'pony-feeding' and 'cattle-feeding' sub-divisions of the grazing communities. The patterns of habitat use described here for both cattle and ponies, restricted though they may be in the case of cattle by constraints of social organisation, may clearly be explained in large part by the animals' requirements from their environment in terms of food and shelter. Seasonal differences in community use reflect seasonal changes in the need to worry about shelter, and also reflect changes in the quality and quantity of forage available in different habitats. This relationship between habitat use and forage availability and quality will be discussed in detail in the next chapter.

Chapter 4

Food and Feeding Behaviour of Domestic Stock

Without a doubt, grazing by domestic animals constitutes at present one of the dominating factors in the ecological functioning of the New Forest system. Grazing has always played an important part in the ecology of the Forest; in the past the deer populations for which the area was primarily set aside perhaps played the major role (populations were estimated at 8,000-9,000 animals during their peak). But, since the decimation of deer populations following the Deer Removal Act of 1851, numbers have never exceeded 2,000-2,500, and from that time on commonable animals have probably been the most significant herbivores.

Coincident with the passing of the Deer Removal Act was the

Table 4.1: Proportions of different grazing animals (expressed in thousands) in the period 1789-1981

Historical period	Deer	Ponies	Cattle	Total grazers
1789-1858	9	3	6	18
1858-1900	3	2.5	3	8.5
1900-1930	2	1.5	2	5.5
1930-1943	1.5	.5	1	3
1943-1965	2.0	1.5	1	6.5
1965-1981	2.5	2.75	1.75	7

Source: Modified from Countryside Commission Report (1984).

increased enclosure of land for silviculture. Although the wild deer are still free to roam over the entire Forest area, the grazing area available for common pasturage was curtailed by such enclosure, and the effective *densities* of grazing stock upon the Forest have thus been further increased.

Over recent years the unenclosed Forest — approximately 20,000 ha of open waste — has been supporting on average some 2,000 cattle and 3,000 ponies (on 1970-80 figures; numbers have now declined somewhat and 1981 figures suggested a total pasturage of only 1,100 cattle and 2,500 ponies). While the area also supports some 2,500 deer (as a maximum estimate), clearly cattle and ponies must have the most significant impact upon the environment just in terms of sheer biomass. Of these two species the ponies probably have the major influence: in terms of numbers, in view of the fact that most of the cattle are pastured on the Forest only during a small part of the year *and* because of the ponies' nutritional physiology. The only non-ruminant among the herbivores, the monogastric horse is adapted to a foraging strategy founded on a very large throughput of but poorly digested forage. This massive throughput, coupled with an ability to crop very much closer to the ground than other species, must make the ponies the species with the greatest single impact upon the Forest vegetation.

In this chapter we shall consider the foraging behaviour and diet of the two main commonable species of the New Forest, the cattle and ponies, and examine the way in which they exploit the varied vegetational resources of the Forest. Seasonal changes in foraging behaviour and dietary composition will be considered in relation to availability and quality of the herbage; we shall also look at differences in the feeding strategies of the cattle and ponies as two different large herbivores living in the same vegetational system.

Foraging Behaviour

The pattern of use of the various habitats of the New Forest by ponies and cattle for feeding is shown in Figures 4.1 and 4.2 which summarise the percentage use of the various community types by ponies and cattle in different seasons.

Results are similar to those presented in the last chapter for patterns of habitat occupancy overall (see also Pratt *et al.* 1985). Feeding use of the various habitat types by *cattle* shows a fairly constant pattern through the year, with heavy emphasis on lawns and improved grasslands and also extensive use of heathland: wet heath in summer, drier areas during the winter months. Feeding use of other communities is not extensive, although deciduous woodland is exploited during the winter, and acid grassland also used for most of the year.

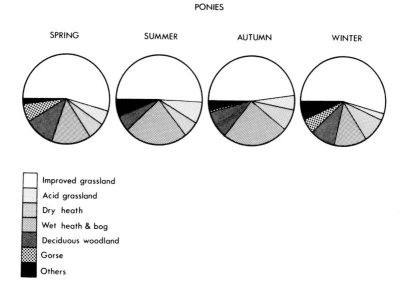

Figure 4.1: Use of different habitats for foraging by New Forest ponies

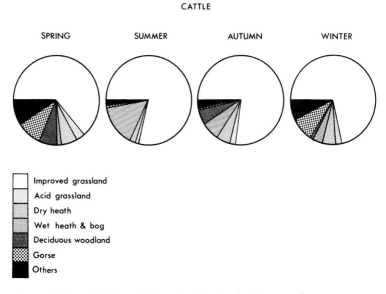

Figure 4.2: Use of different habitats for foraging by Forest cattle

Ponies exhibit a much more marked seasonality in their use of vegetation types. While improved grasslands are still emphasised throughout the year and acid grassland is also used consistently, wet heath, bogs and regenerating heathlands show a clearly seasonal pattern of use, as do deciduous woodland and gorse brake. Feeding use of bogs, while quite high throughout the summer, shows marked peaks of use from April to June and during August and September; use of wet and regenerating heath also shows a clear summer peak between May and September. Use of these latter habitats correlates closely with time of growth of the purple moor-grass (*Molinia caerulea*) in each particular community: *Molinia* is the 'main' forage species taken in bogs, wet and regenerating heaths, and, as will be seen later, is an important element of pony diet at this time, providing abundant forage from these wetter communities even in the drought months of summer when the preferred grasslands dry out and growth is halted. Feeding in gorse brake and deciduous woodland is restricted largely to the winter period, when it may become quite extensive, although woodland feeding continues throughout the year at night, when animals retire to these communities for overnight shelter.

It is quite apparent from these results that feeding by cattle and ponies is not evenly distributed across the different communities of the New Forest. Both species are preferential grazers and for most of the year concentrate the majority of their feeding activity on the various types of improved grassland and on streamside lawns (at all times more than 34% of pony feeding observations and 54% of cattle feeding is concentrated upon these areas, which together total only some 1,460 ha, less than 7% of the Open Forest area). In the spring, growing season (March, April, May) and in the summer (June, July, August), use of grasslands is still higher (ponies 61.5% spring, 47.2% summer; cattle 67.5% spring, 83.4% summer); even outside the main season of production of the Forest grasslands (May-September), feeding occupancy of these communities by both cattle and ponies still remains remarkably high, despite the fact that the standing crop of available forage declines to 25-30g/m^2 towards late summer.

Cattle, indeed, show relatively little flexibility in foraging behaviour throughout the year, and even in midwinter feed extensively on open grasslands. With the decline of available forage on the improved grasslands in autumn, however, the ponies make increasing use of the *Agrostis setacea* of acid grasslands. As winter advances further, they make more use of woodland browse and gorse brake, and this increased feeding use of woodlands and gorse continues through the winter until the first spring growth begins in the more open communities.

Both during the grass-growing season and outside it, cattle make more extensive use of wet and dry heathland than do ponies, and it is primarily an increased use of these communities during the autumn and winter that for them compensates for reduced food availability on grasslands at this time. As noted earlier, however, the majority of cattle are removed from the Forest during the winter. Those that remain are fed, and most feeding sites are concentrated on grassland or heathland areas. Thus, management too may be in part responsible for the apparent lack of flexibility in feeding style.

While this analysis offers information on the relative distribution of feeding animals across the vegetation types of the New Forest, that distribution is, once again, influenced in part by the relative availability — in terms of unit area — of the different vegetation types. If data are corrected for differences in available area of different communities by calculating preference indices of the same form as those presented for habitat use overall (page 53), we may determine more accurately preferred feeding communities. While, for cattle, relatively little selectivity emerges, beyond the gross pattern of occupancy already described with emphasis throughout the year on improved grasslands, ponies show heavy selection for particular communities at different seasons. Among the grassland communities marked preference is demonstrated for streamside lawns, and roadside verges also show consistently higher use per unit area than might be expected. The importance of *Molinia* communities (wet and regenerating heaths and bog) in summer is stressed, while feeding pressure in gorse brake during winter is also extremely high. Other points highlighted are the increased selection for coniferous woodland (probably also deciduous woodland, but available area is so high that the trend is concealed) and woodland glade during the height of summer, presumably as a response to feeding where possible in shade.

Winter use of gorse is one of the most distinctive characteristics of the New Forest ponies. As already noted in Chapter 3, gorse brake is positively selected over the winter months as one of the few communities offering good shelter from deteriorating weather conditions, but it is clear that it is also a preferred feeding community. As a legume, gorse clearly provides a highly nutritious forage at this time (page 75) despite its not inconsiderable defences (Plate 16). The ponies have developed a special technique for dealing with this rather spiny foodstuff: they eat only the tips (which are in any case the most nutritious part) and, when they have selected a shoot, roll back their sensitive upper lip and push the incisors as far forward as possible before carefully biting through the stem. The behaviour is unique to the Forest, and appears to be learnt. It should, however, be noted that not all Forest ponies feed on gorse; use of this foodstuff, and the behaviour required in order to exploit it, seem to be characteristic of particular local populations, and seem to be acquired by individual foals by copying their mothers or others in the immediate social group. There is a considerable local tradition, and *some* objective evidence (Gill 1984), that those ponies which *do* feed on gorse maintain a far better body condition over winter than those which do not have the behaviour.

The link between increased occupancy by ponies of bogs and wet and regenerating heath and feeding use of *Molinia* has been noted. Feeding use of *all* the different communities may indeed prove to be determined by forage availability. The degree of relationship between feeding use of the different grazing communities (in this case: numbers feeding per unit area) and monthly standing crop and productivity of forage has been examined (Putman *et al.* 1981 and 1985) to see if there is some correlation between forage availability and feeding use of an area.

Correlation overall, of feeding occupancy with standing crop, is not close, either for cattle or for ponies. Cattle show slight positive correlation between use of a community and standing crop during spring (March-May) and summer (August-September), but this is not statistically significant. Ponies show a consistent *negative* correlation with standing crop which is significant between May and December. Clearly, feeding occupancy of a community is not influenced by standing crop of available vegetation as we hypothesise; rather, standing crop is reduced in those communities with high feeding activity by the ponies. Relationships with productivity are even less clear, and no consistent correlation between feeding use of a community and its monthly productivity was established (Putman *et al.* 1985).

Such a result conflicts with a clear subjective impression that the timing of animal use of various forage communities is indeed closely marked to productivity, with ponies seeking at any one time the vegetation types showing maximum production. Certainly, at the beginning of the growing season in early spring, community use *is* closely related to production. As soon as the grasses and herbs begin to grow once more, the ponies at once forsake the browse communities in which they have foraged over winter and return to the more open 'grazing' communities. Clearly, they retain a marked preference for grazing rather than browsing and they move back onto these open communities at the earliest possible opportunity. At this time, the pattern of habitat use is closely related to the *onset* of production in each vegetational type. The first community types to flush are the bogs and streamside lawns, and it is noticeable that feeding occupancy of these habitats by ponies is relatively higher during early spring (February/March) than at other times of year (except during the drought months of midsummer when, once more, these wetter communities are the only ones showing significant growth). As the drier grasslands of re-seeded or Commoners' improved areas begin to flush a few weeks later, cattle and ponies begin to move onto these more preferred communities, once again matching their selection of communities to the timing of onset of new growth, until the distribution of both species shifts towards the more uniform pattern of grassland use characteristic of the main growing season.

It is here, however, that the relationship between habitat selection and productivity breaks down, for the factors determining habitat occupancy are complex — and they relate only in part to food availability. As we noted in the last chapter, habitat selection is the end result of a complex interplay of a host of different needs. Thus, in practice, the animals do not in fact select particular communities specifically for foraging. They feed within an overall pattern of habitat use which is determined by all considerations in complement: food, shelter, and whatever else. Early in the season, when palatable food is in short supply, foraging needs may indeed take priority, and a close relationship may obtain between observed patterns of habitat use and forage quality — as noted here, and also reported by Duncan (1983) for horses in the Camargue. But, as the growing season advances and food becomes more readily, and more uniformly, available, other factors — the need for water, cover or 'shade' — may predominate in habitat choice and the animals feed within a pattern of habitat use determined by these other factors.[1]

Patrick Duncan's data on habitat use and activity patterns of Camargue horses (Duncan 1983) offer interesting comparison with patterns reported here for the ponies of the New Forest. Although no woodlands are available to the Camargue horses, and the habitat is

---

[1] One especial consideration here is the purely social phenomenon of 'hefting'. As we have discovered, the ponies are preferential grazers, and indeed must focus their whole home area on a suitable grazing site. So important are these grassland areas, and so strong is the site attachment, that the ponies spend a disproportionately high amount of their time on these favoured lawns. From the time they show the first signs of growth on to well into the late autumn, the ponies heft to these areas, showing remarkable reluctance to leave the lawns even when they are virtually grazed out and other adjacent communities offer acceptable forage in much greater abundance.

overall far wetter than that of the New Forest, many of Duncan's vegetational types have approximate equivalents in the New Forest, and comparisons of patterns of habitat selection reported are quite enlightening in explaining the underlying requirements that the animals are seeking to fulfil from these two very different systems. Duncan reports an increase in use of richer grasslands for feeding from early spring (c. 22%) through to an August peak of 67% use. Feeding use declines through the autumn to lower overwinter levels. The main complement to this pattern of grassland use is reflected in use of heathland areas, which show peak use in early spring (January–March 80%) and a decrease in feeding occupancy over the summer, to 35% in August. Use of coarse grasslands is low (4–7% for most of the year), but peaks over the winter from November to March at approximately 30%.

In the New Forest, use of the coarse acid-grassland vegetation type is relatively constant throughout the year, at the same levels of 4–7% of all feeding observations, but it also shows a slight increase in autumn (August/September) and late spring (April/May). Use of the various improved grasslands and stream banks ranges from around 30% (late winter) to a midsummer peak of nearly 50% (cf. 22%–67%); in both systems ponies are preferentially grassland feeders, and move onto these vegetation types as soon as the first flush appears at the start of the growing season. Use of heathlands and bog communities in the New Forest is, however, nowhere near as extensive as in the Camargue, and shows a reverse pattern of seasonal use: contributing only approximately 20% to total vegetational use from January to March and *increasing* in use through the summer (30% use in August).

This apparent reversal is readily explained. In both the New Forest and the Camargue, grasslands are the favoured feeding areas as soon as forage is available. Over winter, when forage is exhausted, the animals favour other communities. In the Camargue, heathland vegetation types offer the best alternative feeding areas (maintaining the highest standing crop of available vegetation over the winter). In the New Forest, woodland communities are available, offering both shelter and forage; in the Forest, therefore, these are preferred and the more exposed heathlands do not suffer greatly increased use. Use of grasslands in the Camargue increases throughout the summer; in the New Forest, the real peak of use is in early summer, a slight decline in late summer being compensated for by increased use of bogs and wetter heathlands. Once again, the explanation is simple. In the wetter conditions of the Camargue, growth continues throughout the summer on the grasslands; in the New Forest, the grasslands dry out beyond midsummer, growth is suppressed, and more nutritious forage may be found in those wetter communities where growth may be maintained.

Feeding Rates

Rates of feeding in the different communities were recorded as bites per minute during follows of individual animals. Mean values for each month are presented in Tables 4.2 and 4.3. Not all communities will be represented in all months, since feeding use varies.

Bite rates differ markedly between vegetation types, and, within vegetation types, seasonal change is also observed. In a preliminary comparison of feeding rates of ponies on different vegetational communities in November and December, 1979, Putman *et al.* (1981) showed a clear, negative relationship between bite rate per minute and vegetational standing crop (Figure 4.3). A similar relationship is observed *within* communities, with bite-rate changes between months clearly correlated with changes in forage standing crop (and such relationship has previously been shown for both cattle and ponies on *agricultural* pastures by Arnold and Dudzinski (1978)). Such a result could be due to two complementary, but converse, effects. First, if standing crop is very low, bite rate must increase if the animal is to maintain a constant rate of bulk food intake. (In addition, if standing crop is low, chewing time per mouthful is decreased, permitting a higher bite rate.) Secondly, since grazing pressure on the Forest is heavy, a sustainedly high standing crop in any vegetation type can be due only to a high standing crop of non-forage species. Since the animals must therefore take time to select preferred species from among this high biomass of non-forage plants, bite rate will decrease. (In this latter case, bulk intake will also decrease. It is to be noted that, for example, dry-heath communities are rarely fed on by the ponies: time spent feeding in such communities in proportion to their percentage contribution to total area of range is extremely low.)

Such relationship between feeding rate and standing crop may also explain most seasonal changes of feeding rate *within* communities. Pony feeding rate on re-seeded lawns, as an example, increases through the winter, December-April, until the start of the growing season. As standing crop increases (productivity exceeds offtake during the summer months: Putman *et al.* 1981), bite rates decline. At the end of the summer, when productivity falls, and offtake starts to remove 'accumulated capital' in reducing standing crop (August/September),

Table 4.2: Feeding rates of New Forest ponies

Mean bite rate, as bites per minute, in different vegetational types.

Community	Jan	Feb	Mar	Apr	May	June	July	Aug	Sept	Oct	Nov	Dec
Improved grasslands (RL & CI)	57	—	63	63	63	59	54	53	55	61	52	52
SL	—	51	60	57	72	52	49	43	46	61	51	51
AG	51	—	39	42	48	37	34	38	32	37	35	—
DH	—	—	13	17	—	44[a]	44[a]	23[a]	—	20	21	22
WH	—	—	41	—	39	42	36	29	29	34	—	—
Bog	—	33	22	41	40	32	37	31	35	38	31	—
DW	43	46	43	39	51	—	43	43	48	43	48	—
CW	—	—	41	—	—	45	42	—	35	—	33	—
WG	57	—	44	63	63	65	64	60	58	69	63	—
Gorse	8	8	6	—	—	—	—	·	—	—	·	5

Note: [a] Dry-heath figures for June, July and August are for feeding on *Molinia*; April, October, November, and December for feeding on *Calluna*.

Table 4.3: Feeding rates of New Forest cattle

Community	Jan	Feb	Mar	Apr	May	June	July	Aug	Sept	Oct	Nov	Dec
Improved grasslands (RL & CI)	—	64	67	65	70	66	71	77	79	70	71	63
SL	—	57	—	—	—	—	61	58	77	56	76	—
AG	47	—	37	43	44	44	33	51	53	44	73	62
DH	40	38	—	36	30	32	39	44	56	39	41	—
WH	22	—	23	23	26	21	—	54	59	—	46	—
WG	—	—	63	—	—	—	—	—	—	—	—	—

Note: [a] Dry-heath figures for June, July and August are for feeding on *Molinia*; April, October, November, and December for feeding on *Calluna*.

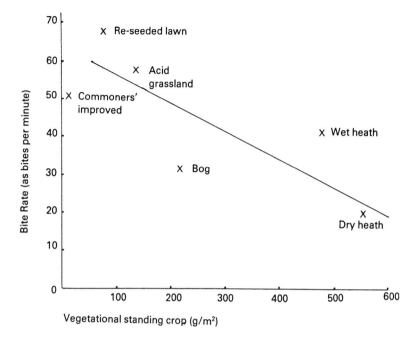

Figure 4.3: Relationship between pony feeding rate and vegetational standing crop

bite rate gradually rises once more. Similar patterns can be observed in other communities.

Diet

The above patterns of feeding behaviour may be more clearly interpreted with a more precise knowledge of actual foods taken. Dietary composition of cattle and ponies may be assessed by standard techniques in analysis of plant residues found in faecal material.

Results presented here for dietary composition of cattle and ponies are derived from microscopic analysis of faecal material. Fresh faecal material, collected monthly during 24-hour follows of individual animals, was analysed by microhistological means, and forage fragments contained were identified by reference to a prepared reference collection of known forage species. Samples of faeces were mixed in water; sub-samples were removed to a petri dish, and scanned with a binocular microscope at magnification × 3000 (methods after Stewart 1967, Stewart and Stewart 1970, and Hansen 1970). Solutions of cow dung were first centrifuged down at 600 rpm to concentrate the larger, more easily identified fragments (Putman 1984; Putman *et al.* 1985).

Tables 4.4 and 4.5 present figures for percentage species composition of identifiable fragments in cattle and pony dung. For both species, separate analyses were carried out for faecal matter collected in two different study areas within the Forest. Dietary composition did not differ significantly between the two areas in any month, and thus all data are bulked for analysis here (see also Figures 4.4 and 4.5).

Cattle

The diet of the cattle remains remarkably similar throughout the year: virtually constant in terms of species composition, varying only in minor changes in relative proportion of those components.

Cattle in both our study areas were fed hay from November to March (as is indeed usual for all cattle on the Forest). Hay and 'fresh' grass were indistinguishable in our analyses: together they made up some 70-75% of the diet throughout these winter months; the balance was made up with heather (both *Calluna* and *Erica*), which comprised a further 20-25% of the total. Other items were rarely more than 1-3% of the diet at this time. There was no significant difference in dietary composition between any of these months.

Diet in April did not differ statistically significantly from that in March, but it is clear that the animals are taking less heather and more grass at this time. Further, although in April 1980 (at the time of our study) hay was fed to a small extent in some areas, the amount available was far less than over the winter months. Thus, had we been able to distinguish between hay and grass in our analysis, differences between March and April would no doubt have been enhanced. That there is a gradual change, however, can be seen quite clearly by comparison of, for example, January and April, where a clear, statistically significant difference is observed in dietary composition, rather than as here between month and succeeding month.

Throughout the summer (May-August) the diet again remains constant, with no differences between months, however paired. Here, about 80% of the diet is composed of grass; heather still provides about 14% of the diet. Nor is any significant difference seen between this summer diet and that of September or October, although there is some

Table 4.4: Percentage composition of the diet of New Forest ponies at different months of the year

Jan	Feb	Mar	Apr	May	June	July	Aug	Sept	Oct	Nov	Dec	
49	37	43	65	90	90	92	87	83	79	69	51	grasses
0	0	0	0	0	0	0	0	0	0	0	0	herbs
0	0	0	0	0	0	0	0	0	0	0	0	conifers
20	26	25	13	0	0	0	0	0	3	11	13	holly
0	0	0	0	2	1	0	1	1	0	0	0	other broadleaves
7	7	5	3	1	1	1	1	2	3	5	10	heather
0	0	0	0	0	0	0	0	0	0	0	0	bramble/rose
0	0	0	0	0	0	0	0	0	0	0	0	ivy
12	13	10	1	0	0	0	0	0	1	3	10	gorse
0	0	0	0	0	0	0	0	0	0	0	0	fruits
12	17	17	18	7	8	7	11	14	14	12	16	other

Table 4.5: Percentage composition of the diet of New Forest cattle at different months of the year

Jan	Feb	Mar	Apr	May	June	July	Aug	Sept	Oct	Nov	Dec	
75	67	71	80	80	83	82	70	66	70	70	66	grasses
0	0	0	0	1	0	0	0	1	1	0	1	herbs
0	0	0	0	0	0	0	0	0	0	0	0	conifers
0	0	0	0	0	0	0	0	0	0	0	0	holly
1	1	1	1	0	0	0	1	2	1	2	1	other broadleaves
21	27	24	9	14	12	13	18	23	19	21	22	heather
0	0	0	0	0	0	0	0	0	0	0	0	bramble/rose
0	0	0	0	0	0		0	0	0	0	0	ivy
0	0	0	0	0	0	0	0	0	0	0	0	gorse
0	0	0	0	0	0	0	0	0	0	0	0	fruits
3	5	4	10	5	5	5	11	8	9	7	10	other

reduction in the percentage of grass taken in autumn, heather intake begins to increase again towards winter levels, and *Molinia* is consistently found in the diet for the first time.

Hay is again fed from November. Without this, no significant difference is recorded in the diet between autumn and winter. Indeed, without the distinction between hay feeding in the winter and grass feeding in the summer months, there is no significant change at all in cattle diet throughout the year, merely a slight change in emphasis between grass/hay and heather.

Ponies

By contrast, pony diet shows marked seasonal change, with characteristic winter and summer forages, and marked spring and autumn 'change-over periods'. Summer diet (May-August), like that of the cattle, is predominantly of grass, which comprises between 80% and 90% of the diet at this time. In contrast with cattle, the ponies feed extensively on *Molinia* at this period: *Molinia* makes up nearly 20% of the total diet, 22% of the total grass intake at this time.

During September and October, *Molinia* intake declines, to only 3% (*Molinia* is a deciduous grass and becomes sere and unpalatable at this time); total grass percentage, however remains relatively constant at about 80%, with a greatly increased intake of *Agrostis setacea* balancing the *Molinia* decline. There is also considerable use of bracken as forage over this period. Overall, the diet may be seen to differ significantly from that over the summer, although this is a progressive change.

As autumn changes into winter, the percentage of grass in the diet declines, to 50% of total intake. Correspondingly, there is a progressive increase, from October right through until February/March, of the amount of gorse and tree leaves (mostly holly) which are taken. The proportion of moss fragments in the faeces also increases over this period (4.5% to 12.7%) and heather intake, too, is increased in winter (mean 6.0%, November-February).

This is a progressive change right through the winter; again, no significant difference is recorded between successive months, but diets of September vs November, October vs December, and November vs January differ significantly. The diet of this winter period also differs significantly from that of autumn and obviously also from that of summer.

At the end of the winter, a 'spring' diet may be determined in the change-over period back to the summer mixture. This diet, with an increase in grass intake and with declining, but still significant levels of

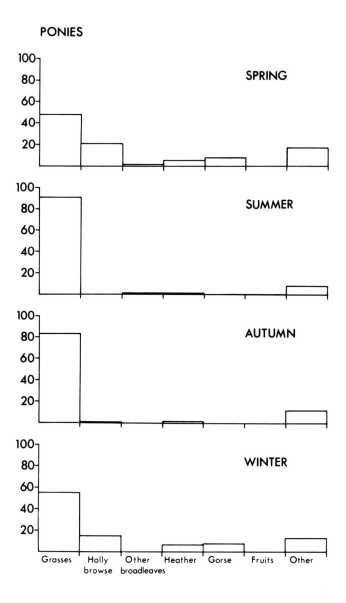

Figure 4.4: Diet of New Forest ponies. The relative proportional contribution to the diet of different foodstuffs (as percentage of total diet) is shown for different seasons

Source: Data from Putman et al. (1985).

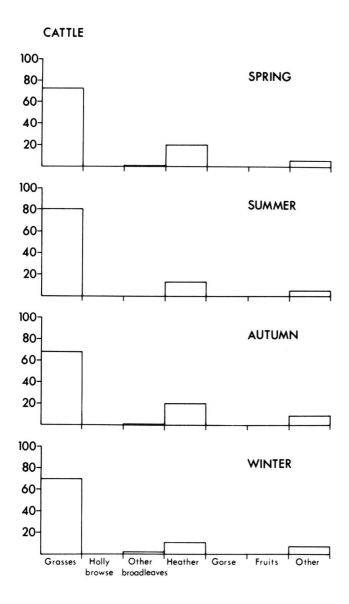

Figure 4.5: Dietary composition of New Forest cattle

gorse, moss and tree leaves, differs significantly from both true winter and summer diets.

Pony diet is clearly much more flexible than that of the cattle, adapting to changes in forage availability and quality with major changes in dietary choice. Even within the grass component of the diet there are striking seasonal changes. Use of *Agrostis tenuis* is greater in the summer than in the winter; *Molinia* is used extensively in the summer but over a very restricted period from May to August, while *Molinia* and *Agrostis setacea* appear to be complete analogues: with major use of *A. setacea* when no *Molinia* is present in the diet, and greatly reduced intake when *Molinia* is being exploited.

Overall, despite the superficial similarity between the diets of these two herbivore species — both are, after all, preferential grazers — dietary composition is shown to be significantly different in all months. As already noted, cattle are fed hay or straw throughout the winter (November-March). Although ponies do use the hay, it does not form so significant a part of their diet as for cattle. Even without this difference, however, diets of ponies and cattle still differ significantly over this period, with cattle taking more grass/hay and heather than do ponies and with ponies taking considerably more tree leaves, moss and gorse (gorse was never recorded in cattle faeces from any area). Even over the summer (May-August), when both species are eating a great deal of grass (cattle about 80%, ponies about 90%), diets of the two species still differ: the ponies are feeding extensively on *Molinia*, with only 50% of the diet made up of other grasses. Through September and October, diets remain distinct: ponies are still feeding on *Molinia* more than cattle, and the increase in tree leaves and gorse is just becoming evident.

The only other comparable information available on the diet and dietary preferences of a free-ranging population of horses is that deriving from Susan Gates's studies of Exmoor ponies (Gates 1982). The dietary patterns she describes are strikingly similar to those presented here for the New Forest animals; both species composition of the basic diet and seasonal changes in the relative emphasis of different forages closely mirror those observed in the New Forest (Figure 4.6). Like the Foresters, Exmoor ponies graze extensively throughout the year, showing strong preferences for *Agrostis* and *Festuca* swards but feeding to a significant extent upon *Molinia* during summer and autumn. When the availability of such grazing declines in the winter, the ponies compensate with increased intake of gorse and heather. Such a pattern of forage use is indeed strikingly similar to that observed for New Forest ponies, and suggests very similar strategies and responses in the two different

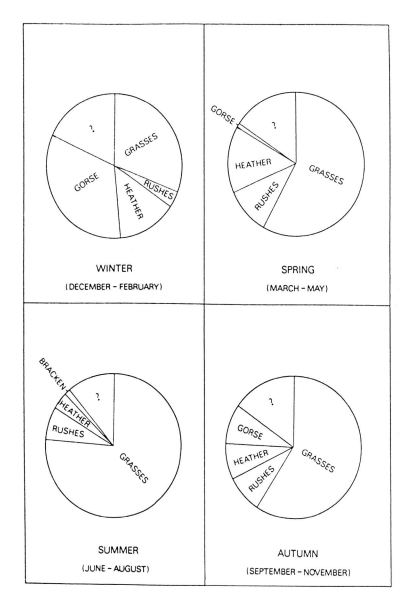

Figure 4.6: Dietary composition of a group of Exmoor ponies for comparison with New Forest animals

Source: Data from Gates (1982).

populations for maintaining themselves on the impoverished grazing of free range.

Diet Composition in Relation to Forage Availability and Forage Quality

Although an analysis of the diet of cattle and ponies ranging on the Open Forest in itself offers interesting information, it is clearly much more useful if we can attempt to *explain* dietary composition and seasonal changes in diet. For, if in so doing we can identify exactly what the animals seek from their vegetational environment at different times, we can perhaps explain the patterns of habitat use discussed earlier — and we approach a more fundamental understanding of the interrelation-ship between grazers and vegetation. In an attempt to identify the factors determining diet at any time, species composition of the diets of both cattle and ponies in each month of the year was related to various parameters of the availability and quality of the various forage species. Dietary composition was tested against standing crop of vegetation (per m²), productivity, calorific value and dry-matter digestibility of the various forage species and the nutrient levels of N, P, K and Ca con-tained (identified as potentially the key nutrients).

Freshly plucked samples of each forage type were collected from the field each month and returned to the laboratory for chemical analyses. Where animals appeared to feed relatively unselectively, samples were taken of the complete sward; in communities where the cattle and ponies selected particular food plants, only these specific species were sampled. Methods used to determine available standing crop of a parti-cular food type or sward, and its productivity of edible forage, are described in more detail in Chapter 7, but these were based essentially on weight yielded over a given time period from regular clipping of sample plots in each vegetation type.

Surprisingly, in neither cattle nor ponies was overall dietary compo-sition found to correlate with any of the vegetational parameters measured. In cattle, such lack of correlation is due to the fact that, as already noted, dietary composition shows little seasonal change in any case and diet is therefore unlikely to be determined by relative quality of different forages. Lack of correlation in ponies is due to the extensive use in winter of tree leaves and gorse, and increased intake of heather: all forages being taken outside their season of maximum productivity or quality. These forages, like the grasses, show their peak productivity and quality during the summer, at which time of year, however, grass is also relatively abundant. The ponies are clearly preferentially grazers and when grazing is available feed predominantly on such forages, turning to browse only when grazing runs short. As a result, their use of browse correlates with lack of availability of grass, rather than peak availability or quality of the browse itself.

If, therefore, browse is excluded from our analysis, and composition of 'graze' species in the diet is related to availability and quality of those species, we find that the relative proportion of dietary intake by ponies of graze forages *does* correlate weakly with relative *productivity* and *digestible nitrogen content* of those forages. Rank order of species in the diet within each month correlates with rank order of productivity and digestible N.* This correlation holds only during the growing season between April and September; no correlation is observed with other vegetational parameters.

In summary, therefore, cattle diet shows no marked seasonal change and is not correlated with any vegetational measure recorded in our survey. Ponies are seen to be preferential grazers, turning to browse in winter when all available grass has been used. Within the grazing component of the diet, use of different forage species is related to their productivity, digestibility and digestible N content — except in the case of *A. setacea,* which once again is taken in any great amount only when other grasses run short.

Total Intake

Once dietary composition has been established, it is theoretically possible to determine not only species composition, but actual intake by weight. Percentage species composition of plant remains in the faeces may be multiplied by total weight of faeces produced during the day, to offer a figure for total weight of faecal fragments of each forage species. If this is further corrected for digestibility of each forage species, we arrive at a figure for total daily intake, by weight, of each of the various dietary components.

Such a calculation makes a number of gross assumptions and compounds a number of sources of error.

(i)   Figures for total faecal output per day are themselves subject to a

---

* These two vegetational parameters are not necessarily independent of each other.

number of errors. Such figures are derived by multiplication of mean faecal weight per defaecation (in itself probably quite accurate) by number of defaecations per 24 hours. It is, however, possible to collect data on faecal events (weights, or times and number of defaecations) only during daylight. Our figures for number of defaecations per 24 hours are thus derived as mean number of defaecations observed per hour x 24, and clearly make the assumption that defaecation is regular throughout the 24-hour period.

(ii) In calculation of total daily faecal output, by weight, of plant remains by combination of faecal weight with percentage composition — by number — of plant fragments, we make the implicit assumption that all fragments are of equal weight.

Such are these assumptions that it would be misleading to present here results for total intake of individual forages. The possible errors discussed here will, however, *not* markedly influence summary values for total nutrient intake. Absolute values presented for daily nutrient intake are associated of course with considerable standard error, but the general trends are quite clear (Table 4.6).

Total dry-matter intake by *cattle* is lowest during late winter, rising through spring to a peak in the summer of 9.6 kg/day. Levels decline once more through late summer and autumn, to return to winter levels of 3.5-4.0 kg/day. Intake reflects, in part, relative availability of forage. Since the cattle do not change their diet significantly through the year, they are unable to compensate for decreased availability of a particular forage by switching to other foodstuff. Thus, in winter, when availability of their chosen forage is reduced, dietary intake must fall, while in summer, when forage is relatively more plentiful, dry-matter intake may be higher. Winter decline in dry-matter intake is, however, probably also due to physiological inappetence, a phenomenon characteristic of all free-ranging ruminants over the winter months whereby voluntary dry-matter intake is reduced considerably below levels of intake during the rest of the year (e.g. McEwen *et al.* 1957; Short *et al.* 1969; Kay 1979; Kay and Suttie 1980).

Associated with this pattern of dry-matter intake, actual *digestible* dry matter ingested also rises in summer and falls in winter. The relatively high digestibility of the grazing in summer, and low digestibility of both grass and heather during the winter, exaggerate the trend, however, and digestible dry-matter intake at its peak in midsummer (4.1 kg/day) is a factor of some five times that calculated for late winter.

Table 4.6: Dietary intake summary: daily intake of nutrients

| | PONIES | | | | CATTLE | | | |
Month	Total dry-matter intake (g)	Digestible dry matter (g)	Digestible protein (g)	Energy (KJ)	Total dry-matter intake (g)	Digestible dry matter (g)	Digestible protein (g)	Energy (KJ)
Jan	5,079	1,991	236	91,176	*4,016	965	110	71,480
Feb	7,031	3,090	339	125,931	*3,840	1,934	84	70,148
March	4,478	1,712	191	81,800	*3,560	840	77	65,746
April	5,157	1,623	213	88,821	*6,183	1,591	224	111,750
May	6,104	2,612	449	105,690	9,657	4,010	536	179,930
June	5,993	2,483	387	106,205	9,613	4,121	507	178,447
July	5,819	2,372	327	104,376	7,191	3,148	357	133,795
Aug	5,836	2,158	336	105,820	7,745	2,953	350	147,060
Sept	6,978	2,312	349	127,720	7,795	2,608	301	150,122
Oct	7,237	2,419	313	136,111	4,846	1,600	165	92,660
Nov	7,858	2,837	350	139,225	*5,848	1,902	212	109,552
Dec	5,611	2,022	237	99,112	*5,029	1,476	165	91,264

*See text, page 82.

It should be noted, however, that these figures have been calculated using digestibility values for grass, during winter, for the entire grass-like forage (grass and hay) intake. The feeding of hay and straw is, however, extensive over the winter period, as we have already mentioned, and, if we recalculated our figures using the digestibility values for hay over this period for the grass/hay component, values for total dry-matter intake and digestible dry-matter intake rise to new totals as

	Nov	Dec	Jan	Feb	March	April
Dry-matter intake (g/day)	6,271	7,181	6,731	5,790	6,059	10,573
Digestible dry matter (g/day)	3,490	2,872	2,701	2,362	2,376	3,943

with both dry-matter intake and digestible dry matter still somewhat below summer levels, but digestible dry-matter levels now maintained at approximately 2.4 kg/day. Such assumption will of course offer an outside maximum, since much of the supplementary forage offered, as straw, has a far lower nutrient value; the true values probably lie somewhere between these two sets of figures.[3]

Total dry-matter intake and digestible dry-matter intake by *ponies* remain remarkably consistent through the year. By changing their diet, as already discussed, the animals are able to maintain a dry-matter intake of between about 5.0 kg and 7.5 kg per day throughout the year. Only in March and April (when the animals start to forsake tree leaves and gorse for, relatively poor, grazing) does daily dry-matter intake fall

---

[3] Correction of total intake figures for cattle by using values for hay rather than grass during the winter months also changes figures for digestible protein and total energy intake, which become

	Nov	Dec	Jan	Feb	March	April
Digestible protein (g/day)	277	250	226	197	202	378
Energy (kJ/day)	136,130	121,436	107,345	99,646	98,010	160,585

below 5.0 kg. Digestible dry-matter intake also remains relatively high throughout the year (2.0-3.0 kg/day; 1.6-1.7 kg/day in March and April), while the remarkably constant ratio which is maintained between digestible dry matter and total intake highlights the ability of the animals to adapt their diet at any particular time of year towards an appropriate range of digestible forages.

Although these calculations for total forage intake are clearly some-what speculative, and compound a number of initial assumptions, the possible errors should not affect results too markedly if these are expressed in terms of total dry-matter intake or total intake of particular nutrients. Further, changes in relative values between months will reflect real difference in forage intake.

It is clear from Table 4.6 that, while there is considerable seasonal variation in total dry-matter intake by ponies, intake of digestible dry matter is maintained at a relatively constant level throughout the year (although when digestibilities of all forages fall to their lowest levels at the end of winter — March/April — total intake of digestible dry matter does start to decline slightly). Cattle show far wider fluctuations in total intake of both digestible dry matter and protein. With a much less flexible feeding strategy/much more constant dietary composition, they are less able to compensate for changing quality/availability of their staple foodstuffs. When availability, digestibility and nutrient status of grasses fall over autumn and winter, cattle total intake also declines dramatically until, through January-March, they must be feeding at well below maintenance.

Feeding Behaviour of Cattle and Ponies: different strategies of exploitation

One of the most striking features of all these data is this marked difference in the flexibility of patterns of habitat use and diet, over the course of the year, between cattle and ponies. While both species are preferential grazers, feeding on the improved grasslands of the Forest so long as forage is available, cattle show little flexibility of habitat use — or of diet — and maintain essentially the same pattern of feeding behaviour throughout the year. By contrast, ponies show pronounced adaptation to changing circumstances. Although they, too, linger on the favoured grasslands for some time after production is exhausted, they eventually drift towards exploitation of other forages during autumn and winter, when food becomes scarce. Increased intake of gorse and other browse (notably holly) from November to January is a striking feature of pony diet (Table 4.4).

It is difficult to account fully for this pronounced difference between the two species. In part, the greater apparent versatility of ponies may simply reflect greater physiological specialisation for grass feeding in cattle. Although both species are *preferential* grazers, the direct-throughput system of the monogastric horse does enable it, if *necessary,* to cope with alternative foodstuffs even if of low digestibility; by contrast, the whole design of the cattle rumeno-reticulum is geared for bulk fermentation of grasses (Hofmann 1982) and perhaps does not permit it to exploit efficiently a wider range of forage. The difference in foraging pattern between the two species is, however, also related to differing social structures. The basic social unit of ponies within the New Forest is the individual — a single mare, or mare plus current foal (Chapter 3). These individuals drift independently around their home range, moving freely from one vegetation type to another. The cattle by contrast are strongly herding, and groups of 30-40 animals move as a unit around a defined range. The pattern of range use is almost stylised, with each herd following a fixed circuit each day, moving purposefully from one vegetation type to another, from night-time shelter in the deciduous woodlands to daytime grazing areas. Further, social cohesion is such that they are unwilling to occupy vegetational communities whose extent is less than about 10 ha, since on smaller areas it may be impracticable to accommodate the entire herd; as a result, many of the vegetation types of the Forest which occur only in small patches are socially unavailable to them: the herd is restricted to woodlands, grasslands and heathlands.

During the summer, growing season, the pattern of habitat use for feeding and the diet are remarkably similar for the two herbivore species. Both concentrate their feeding upon the various improved grasslands of the Forest, with peripheral use of wet and dry heathland, acid grassland (predominantly ponies), bogs and deciduous woodland. In both species, dietary composition at this time is over 80% grass. Are they in competition at this time?

Closer investigation of the data reveals that, although feeding use of grasslands by both species is extremely high at this time, the two herbivores concentrate on *different* grassland types. Such separation is in fact apparent throughout the year and not just in the season of highest apparent overlap. Thus, by way of example, cattle tend to concentrate most of their feeding activity throughout the year on re-seeded lawns; Commoners' improved grasslands are exploited during March, April and May (the peak growing season), while streamside lawns and road-side verges are used at far lower intensity. Ponies feed more evenly over all grassland types, but none the less show a preferred use of Common-ers' improved grassland for much of the year (particularly from October to January/February), more consistent use of streamside lawns and relatively high use of roadside verges. Two other grassland types surveyed in this way (acid grasslands and the *Agrostis tenuis* glades of woodland clearings) are used virtually exclusively by ponies through-out the year. Figures for animal density (feeding animals per unit area) show even stronger preferences by cattle for re-seeded lawns, and by ponies for streamsides, road verges and Commoners' improved grass-land.

These results indicate some separation in feeding use of the different grassland communities by the two herbivores, with some spatial separa-tion (cattle do not use woodland glades) and some difference in the importance of shared communities to the two species.

Further *dietary* separation is also evident from Tables 4.4 and 4.5. Throughout the summer months, ponies make extensive use of the deciduous grass *Molinia caerulea*, gleaned from acid grasslands, bogs and heathlands around the edges of improved grasslands. Between May and August, *Molinia* may comprise 20% of the total diet, 22% of grass intake of ponies. This high *Molinia* intake coincides with a time when the rest of the diet most closely approaches that of cattle, and perhaps preserves a separation between the two species at the most critical time. Cattle were observed feeding on *Molinia.*

The degree of overlap between the two species in terms both of habitat use and of diet may be more formally examined in calculation of actual 'niche overlap'. The niche-overlap index of Pianka (1973) examines objectively the extent to which a number of species overlap in their use of a particular ecological resource (food, habitat, time, etc.). It calculates for each species what proportion of its needs in a given resource is met by different parts of that resource and then examines the extent to which this overlaps the pattern of use of those same resources by the other species (see e.g. Pianka 1973, Putman and Wratten 1984). The index may assume values from 0 (total ecological separation) to 1 (total overlap). Calculated for our New Forest data in terms of diet and habitat use, these indices allow a more objective assessment of overlap. Surprisingly, they suggest considerable dietary overlap throughout the year. Despite the extensive use made by ponies of *Molinia* during the summer and of holly and gorse during the winter months, and despite the extensive use of heather by cattle, calculated indices of overlap

Table 4.7: Niche overlap between cattle and ponies in the New Forest in use of habitat and in diet

	Spring	Summer	Autumn	Winter	Whole year
Overlap in habitat use	0.89	0.80	0.65	0.59	0.78
Overlap in dietary composition	0.87	0.95	0.96	0.93	0.96
Combined overlap	0.77	0.76	0.62	0.55	0.75

Table 4.8: Niche overlap between cattle and ponies in the New Forest

	Diet	Habitat use	Time	Combined overlap
Spring	0.87	0.89	0.82	0.63
Summer	0.95	0.80	0.92	0.70
Autumn	0.96	0.65	0.93	0.58
Winter	0.93	0.59	0.77	0.42
All months combined	0.96	0.78	0.93	0.70

Overlap as: $\dfrac{\sum p_{ia} p_{ja}}{[\,(\sum p_{ia}^2)\,(\sum p_{ja}^2)\,]^{\frac{1}{2}}}$ (Pianka, 1973)

never drop below 0.8 in any month. Overlap in habitat use, however, is high only during the first flush of growth of early summer, being 0.6 or below for most of the rest of the year. The actual ecological separation achieved in practice by both these separate effects acting in combination may be derived as the product of the two separate indices (Table 4.7). On theoretical grounds it has been calculated that, where overlap is restricted to a level less than 0.54, the animals are unlikely to be experiencing severe competition (MacArthur and Levins 1967). From Table 4.7 we may deduce that the animals risk competition only through the growing season between March and July. Further, actual overlap is probably considerably lower than is apparent here, for a number of other factors may contribute to effective separation in practice. The time of day when the animals exploit the different habitats and food resources may be crucial in avoiding competition, and, although we may also take this into account (Table 4.8; Putman 1985b), we are still not allowing fully for observed spatial or 'geographical' separation. While cattle and ponies may both make extensive use of improved grasslands, they do not necessarily use the same areas, or the same areas within them. As we noted in the last chapter, cattle rarely make use of grasslands whose total area is less than about 10 ha. In addition, where both species *do* occur together on the same favoured grasslands they occupy quite distinct and separate patches within them (page 58-9). In practice, therefore, quite a high degree of separation may be achieved between the two species.

Chapter 5
Ecology and Behaviour of the Forest Deer

Fallow Deer

Fallow deer (*Dama dama*) have long been the dominant deer species in the New Forest; indeed, William I's declaration of the area as a Royal Forest was chiefly to conserve hunting interests for this species. It is difficult to assess what numbers may have occurred on the Forest at that time; the earliest reliable census is that of 1670, when the Regarders gave a return of 7,593 fallow deer and 357 red deer within the Forest boundaries. A government report of 1789 gives an average number of fallow deer present each year as 5,900, and numbers seem to have been maintained at roughly this level until the 1850s. In 1851, the New Forest Deer Removal Act provided for the 'removal' of deer from the Forest within three years of the enactment. Total extermination of such a large population of animals, scattered over so large an area, proved of course totally impracticable, but numbers were certainly dramatically reduced and population estimates in 1900 (Lascelles 1915) gave a figure of 200 head. From then on the population has gradually expanded, and is now maintained by the Forestry Commission at a level which has been estimated at about 2,000 animals (Strange 1976). The deer are free to wander over the entire Forest area, within and outside Inclosures, though most of the population are concentrated within enclosed woodlands.

Fallow are among the most widespread and best-known of all the British deer species perhaps because they are the species most commonly maintained in park herds: their delicate build, spotted coat, and the broad palmated antlers of the bucks make them an attractive decoration for many a private estate (Plate 2). This same interest by man has indeed had a dramatic influence on their world distribution. Although widespread throughout Europe some hundred thousand years ago, the species probably became extinct in the last glaciation except in a few small refuges in southern Europe. From these relict populations, reintroduction to the rest of Europe and to Britain must have been at least assisted by man; it is thought that fallow were brought to the British Isles by the Romans or the Normans (Chapman and Chapman 1976).

There should be little need to go into too much detail at this point on the general biology of fallow deer. The species has been extensively studied and excellent reviews of the general biology have been published by Cadman (1966) and, more recently, Chapman and Chapman (1976). In the present context, therefore, a brief review only will be offered merely to act as a general outline to emphasise the major points.

In the wild state, fallow deer are characteristic of mature woodlands; although the deer will colonise coniferous plantations (provided these contain some open areas), they prefer deciduous woodland with established understorey. These woodlands need not necessarily be very large, because they are used primarily only for shelter: fallow are preferential grazers and, while in larger woodlands they may feed on grassy rides or on ground vegetation between the trees (Plate 11), they will as frequently leave the trees when feeding to invade agricultural or other open land outside. As a result, although the best known populations in England are associated with larger forests such as Epping Forest, the Forest of Dean, Cannock Chase and the New Forest, fallow deer are equally at home in smaller woodlands or scattered copses in agricultural areas.

Traditionally regarded as a herding species, fallow have rather a complex social system and social organisation is closely linked to the annual cycle (Figure 5.1). In larger woodlands males and females remain separate for much of the year, with adult males observed together in bachelor groups and females and young (including males to 18 months of age) forming separate herds, often in distinct geographical areas. Males come into the female areas early in autumn to breed; mature bucks compete to establish display grounds in traditional areas within woodland, and then call to attract females. From this time on, adult groups of mixed sex may be observed through to early winter. Rutting groups then break up and the animals drift away to re-establish single-sex herds (Figure 5.1).

This picture is, however, somewhat oversimplified. Most of the detailed research undertaken on fallow deer has been done in large blocks of woodland such as the New Forest (Cadman 1966; Jackson 1974, 1977) and Epping (Chapman and Chapman 1976), and results may be influenced by this unintended sample bias. In common with that

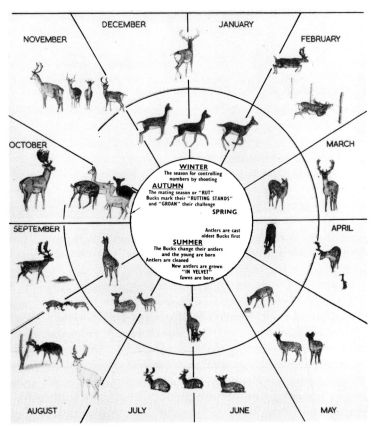

Figure 5.1: The annual cycle of life for fallow does (inner ring of diagram) and fallow bucks (outer ring)

Source: Modified from Cadman (1966). Crown copyright, reproduced by permission of the Forestry Commission.

of many ungulates, fallow deer social organisation seems to be extremely flexible and strongly influenced by environment. Group size is clearly correlated with habitat (page 100) and it appears that relations between the sexes may also differ in different environments. Thus, while in the large areas of relatively continuous woodland in which they have normally been studied, the sexes do indeed remain separate throughout the year, except for a brief and well-defined rut, elsewhere different patterns may be observed. In agricultural areas for example where small woodlands are scattered amongst arable land, group sizes tend to be larger and males remain longer in doe areas after the rut is over, with the doe groups right up to the end of April or May. These 'harem' groups do eventually break up as individual does move away when the fawning period approaches; the bucks return to bachelor groups after antlers have been shed and then remain in these male-only herds until the following rut. In other areas again, however, — usually in largely *open* habitats — mixed herds containing adults of both sexes may persist throughout the year.

Various evolutionary theories have been established to explain the tendency of all deer species to adopt single-sexed herds for part or much of the year (e.g. Clutton-Brock and Albon 1978; Staines *et al.* 1981; van Wieren 1979). Yet these cannot account for this complete continuum of strategies among fallow deer populations. We have already suggested that group size in New Forest ponies might be affected by vegetational disposition, and by size-scale of the environment (O'Bryan *et al.* 1980). I consider it likely that similar factors are influencing herd structure in fallow deer. Thus, I believe that 'single-sexing' may occur in large-scale environments — large in terms of overall available area as well as vegetational mosaic — and may be related also to the more closed nature of woodland habitat. Persistent harem groups and totally mixed populations are more characteristic of 'island' woodlands and enclosed areas such as deer parks, and may be linked both to a small-scale vegetational mosaic and to the more 'restricted' nature of the environment — which offers little opportunity to 'get away' and establish separate sex groups Further, it seems possible that the degree of sex separation may depend as much on total population density as on such vegetational cues. (A small population subdivided into component sexes might produce herds too small to be separately viable.) These explanations are, however, mere speculation, and within the large-scale environment provided by the New Forest (whether it is typical or not!) males and females do operate as independent sexes outside the brief rutting period.

Breeding is highly seasonal. Although does may be receptive at any time between September and March, and mature bucks may have active sperm in the epididymis from July or August, most conceptions are synchronised in the brief, hectic period of the rut. The timing of the rut varies from area to area; in the New Forest, it usually starts towards the last week of September and is at its peak through October; few bucks call beyond the first two weeks of November. Rutting stands are very traditional and the same buck may hold the same display ground for a number of years. Does, too, are faithful to a particular rutting stand, and will return to the area each year, frequently accompanied by their daughters. (Fallow does do not disperse widely from their natal range, and females often establish ranges which overlap extensively with that of their mother.) As a result, a buck may commonly cover his own daughters — even his grand-daughters: the implications of such regular inbreeding are considered elsewhere by Smith (1979). Does give birth to single fawns in late June or July, eight months or so after conception. For a number of weeks before and immediately after the birth, they become very independent and secretive in habit. Groups re-form only when the fawns are already a few weeks old. At this same time of year bucks, too, are somewhat solitary. Antlers are shed in April/May, and from that time the males keep themselves rather to themselves until the new antlers are quite well grown (Figure 5.1). Bachelor groups then re-form until mature animals leave the buck areas to establish rutting stands in the autumn.

Rutting stands are not always held by a single male: although mature bucks defend their stands aggressively against most potential rivals, they quite frequently tolerate the continued presence of one other, particular, male on their 'patch' (often, but not necessarily a *younger* buck). Such 'satellite' males are by no means uncommon: male deer frequently form close bonds with one other individual in their bachelor group — what one might almost describe as friendships — and where rutting stands are shared it is always between males which already have this close social bond. Most rutting stands, whether held and defended by a single male, or a pair of 'friends', are usually widely separated from each other. Occasionally, however, a whole system of stands may be established very close to each other, with as many as eight or nine adjacent rutting stands in an area of only a few hundred square metres. Such 'multiple stands' have been reported in two main areas within the New Forest, but they are not unique to this area, for they are also recorded by Chapman (1984) in Essex. The reasons for this occasional 'aggregation' of rutting stands are unclear: perhaps the does in these areas are at particularly high densities, or suitable habitat for display grounds is scarce in that particular area. Whatever the cause, competition for females in these areas must be extremely severe, and these multiple stands present a system of communal display almost analagous to the leks of birds such as ruff, blackgrouse and capercaillie.

The above review presents, with a rather broad brush stroke, a simple picture of the biology of fallow. Within this general framework, let us now turn to examine in more detail the biology of fallow deer within the New Forest itself.

Plate 1 A group of Forest ponies (*R.J. Putman*)

Plate 2 New Forest fallow bucks (*R.J. Putman*)

Plate 3 Sika hinds in summer coat (*R.J. Putman*)

Plate 4 Grazing roebuck (*R.J. Putman*)

Plate 5 New Forest ponies 'shading' (*Stephanie Tyler*)

Plate 6 Acid grassland merges into the richer grazing of streamside lawns on the alluvial soils of river floodplains (*R.J. Putman*)

Plate 7 Improved grassland (*R.J. Putman*)

Plate 8 The *Calluna*-dominated dry-heath community (*R.J. Putman*)

Plate 9 Wet heath and bogs occur wherever drainage is impeded (*R.J. Putman*)

Plate 10 Deciduous woodlands of the Open Forest have little ground vegetation or shrub
(*R.J. Putman*)

Plate 11 A group of fallow deer in the New Forest clearing (*R.J. Putman*)

Plate 12 Catching fallow deer in the New Forest to fit radio-transmitter collars (*Timothy Johnson*)

Plate 13 Temporary exclosures for measuring vegetational productivity in place on a streamside lawn (*R.J. Putman*)

Plate 14 Forest grasslands are cropped extremely close by the grazing ponies (*R.J. Putman*)

Plate 15 Browsing results in a lack of shrub layer within woodlands and a distinct 'browseline' on the woodland trees (*R.J. Putman*)

Plate 16 Gorse bushes are 'hedged' by continuous browsing pressure (*R.J. Putman*)

Habitat Use

The pattern of use by fallow deer of the various habitats offered by the New Forest has been studied by Jackson (1974) and by Parfitt (1985). Both concentrated their attention on populations based within Inclosures, rather than on the Open Forest; the deer thus had available to them a rather restricted array of habitats. Of those communities recognised in our studies of cattle and ponies (Chapter 3), Parfitt and Jackson consider that the deer had ready access only to woodland habitats (deciduous woodland, mature softwood stands, plantations, prethicket and thicket, rides and glades). Both authors describe a similar pattern of use of these available habitats (e.g. Table 5.1). In each case, the observed number of animals in each habitat differs significantly from the expected distribution if use of the range of vegetation types was uniform. Deciduous woodland was actively selected in early spring (February-April) and autumn (August-October). Woodland use remained high throughout the winter in good mast years as animals remained to feed on the abundant beech mast and acorns. Where mast was less abundant, use of the woodland blocks themselves declined over the winter and the deer made increasing use of more open habitats, grazing along rides and in clearings; rides and clearings were again used heavily in midsummer (June/July). As with cattle and ponies (Chapter 3), changes in habitat use reflect changes in the need for shelter and in the availability of favoured foodstuffs.

As noted, both Jackson and Parfitt consider the use of habitat primarily of fallow deer within Forest Inclosures. It is, however, clear that many fallow deer *do* forage out onto the Open Forest. Jackson

Table 5.1: Habitat use by New Forest fallow deer (Parfitt 1985)

	Percentage of observations in any one month recorded in each available habitat											
	J	F	M	A	M	J	J	A	S	O	N	D
RL/CI SL	not available in environment											
AG/WH	9	4	2	0	2	9	2	2	1	2	9	9
DH	0	0	0	0	0	0	0	0	0	0	0	0
Bog	0	0	0	0	0	0	0	0	0	0	0	0
DW	13	49	58	43	16	20	6	46	46	40	20	26
CW	35	38	8	32	38	33	23	28	28	26	19	23
WG and Rides	31	25	19	20	30	30	59	26	26	28	42	38

presents an analysis of habitat use by these deer as in Table 5.2. Deer near the Forest edge or whose range abuts agricultural land within the Forest may also regularly graze out on agricultural pasture, increasing their effective use of grasslands; the degree to which use of agricultural land changes the pattern of habitat use overall is, however, uncertain.

Home Range

Individual animals restrict their activities to particular home ranges (although these overlap considerably and the animals are in no way territorial). In a superb study of the social behaviour of New Forest fallow deer, Norman Rand (unpublished) has collected thousands of sightings of over 600 individually recognised animals. Rand has developed a method of identifying fallow deer from the pattern of white spots on the coats of individual animals. Although the pattern of spots on the whole coat can be used for identification, Rand has shown that animals can be distinguished just by the pattern of spots on the thigh. These spots are always larger and clearer than those elsewhere and usually occur in easily recognisable lines or groups (see Plates 2, 11 and 12). Spot patterns are, however, distinct between right and left thigh on any one individual; to avoid confusion, Rand has developed a photofile of right and left sides of all animals recognised. The pattern of spots is constant for one individual, and consistent from year to year. The spots become less clear in the winter coat, but Rand has shown definitively that each individual retains the same pattern through moults from year to year.

Since 1977, Rand has maintained a research programme within an area of some 11 km^2 in the New Forest. Using his individual-recognition

Table 5.2: Habitat use by Open-Forest fallow deer (Jackson 1974)

| | Percentage observations in any one month recorded in each available habitat | | | | | |
	Jan/Feb	Mar/Apr	May/June	July/Aug	Sept/Oct	Nov/Dec
RL/Cl	0	58	68	16	56	0
SL	0	11	5	16	18	0
AG/WH	0	3	0	0	0	0
DH	20	18	0	9	0	0
Bog	0	0	0	0	0	0
DW	80	0	27	50	27	100
CW	0	0	0	0	0	0
WG and Rides	habitat not available					

system, he has steadily accumulated information about the deer normally resident within that area. He has always intended to base his final conclusions on an analysis of the behaviour, associations etc., of known individuals throughout their whole lives. Inevitably, this is a long-term project and he is not prepared to publish prematurely. In 1981, however, he calculated some statistics for range size. For completeness he has agreed to allow inclusion of these previously unpublished statistics in this book, but he wishes me to emphasise his own reservations about their value. They resulted from a relatively brief, preliminary examination of only a small part of the data collected in the years 1977-81. He has found fallow deer ranging behaviour to be highly complex, even intricate. As well as individual variations in range size, there are considerable variations between the years; included in the latter are long-term changes apparently of a sequential nature (there may be a cyclical element), known to be affecting considerably more than the immediate research area. When a full analysis of all the information now available has been completed, Rand believes that it will be necessary radically to revise these statistics.

With these reservations in mind, then, we may none the less offer some provisional estimates for range sizes of fallow deer within the New Forest, based on Rand's preliminary analysis of the 1977-81 data for some 60-70 does and 150 bucks (Figure 5.2). Fallow does in his study area were clearly identifiable only between May and October each year, but over this period full range size varied between individuals from 48 ha to 89 ha (with a mean of 69.2 ha). If a core range is defined on the basis of inclusion of 80% of all sightings, range is defined as 17-46 ha, with a mean of 36.9 ha.

Range sizes of bucks can be determined over a wider range of seasons. Over the same summer period (May-September, excluding records for October when the animals move to traditional rutting stands to breed) full range size had a mean of 107.3 ha with mean core range of 34.9 ha. During winter (November-February) range sizes increased: full ranges varied from 55 ha to 260 ha (mean 152.9), while individual core ranges were between 11 ha and 125 ha (mean 63.3). As foraging conditions improve again in the spring, ranges are reduced (mean 137.7 ha, in March/April (full range); 46.8 ha, 80% range).

Rand's estimates of range size can be criticised on methodological grounds: that the accuracy of identification of individuals based on their coat patterns cannot be guaranteed, and that observations may be biased towards those parts within a range where the animal is more easily visible to an observer. Having examined his data myself in

considerable detail, I am left in no doubt of the authenticity of his identification; Rand's own reservations about the validity of range estimates
quoted here are based on the fact that they are calculated from a relatively small sample of animals and over a four-year period only (while he
himself is already aware of considerable individual variation and year-
to-year change in range size). The inaccuracy of estimating home range
on visual observations is a more difficult criticism to counter, but from
1979 to 1982 an independent study of the range size of New Forest fallow does was undertaken by Andrew Parfitt (Parfitt 1985) using
radiotelemetry. A number of does were captured and fitted with miniature radio-transmitters on plastic collars (Plate 12); after release, the
animals could be tracked by following the radio signal even when they
were not visible to the observer. Parfitt's study was undertaken in a
slightly different area of the Forest from that of Rand, and of course his
results are based on a smaller sample of individuals and a shorter time
period: 35 does were collared between 1979 and 1981. Results from this
work have still to be fully analysed but will make interesting comparison
with Rand's preliminary figures (Parfitt 1985).

Social Organisation

Fallow deer are traditionally regarded as one of the more social of the
deer species, and certainly they may frequently be encountered in large
aggregations. Such sociality, however, is in practice somewhat illusory,
and derives at least in part from animals which occupy overlapping
home ranges coming together in favoured areas to feed.

As noted on page 89, the separate sexes in New Forest fallow deer
typically operate independently of each other for the greater part of the
year, coming together only really for the brief period of the rut in late
September or October. For the rest of the year we may distinguish
between 'bachelor' individuals or groups, and groups of 'small-deer'
(does, fawns and followers), which may even segregate in terms of
geographical area: within any forest large enough to permit such separation one may recognise distinct buck and doe areas. Herds
are formed by individuals from overlapping home ranges coming
together, and clear seasonal changes in size and composition of such
herds may be seen to be closely related to the annual cycle (Figure 5.1).
Figures 5.3 and 5.4 show the frequency of occurrence of groups of particular size among male deer and small-deer in John Jackson's study
areas, and seasonal changes in the frequency with which a particular
group size might be encountered (Jackson 1974). Almost identical

Figure 5.2: Home-range size in fallow deer. Range boundaries may be plotted as a series of 'contour maps', enclosing increasing numbers of animal observations. In the diagram below, the outer line in each case represents the range boundary enclosing all observations of a particular animal, the inner boundary takes in only the innermost 80% of sightings

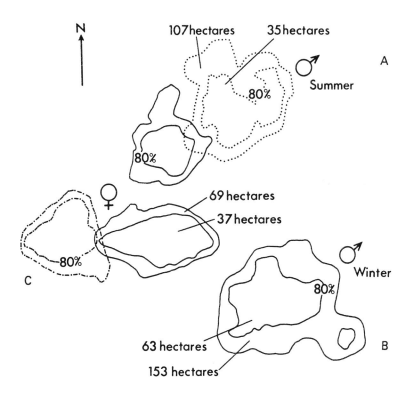

Data superimposed here are mean range areas, from Rand (1981), of
A: ♂ deer; May-September
B: ♂ deer; November-February
C: ♀ deer; May-October

patterns are reported by Parfitt for the deer in his two main study areas; for small-deer herds, Parfitt also plots the change in *mean* group size (Figure 5.5). For both sexes, groups of between one and five animals are most common throughout the year. From November to January, almost all males are encountered singly or in pairs; in February and March, groups of three to five are almost equally commonly observed, but from then on these larger groupings become progressively less and less frequent throughout the summer and autumn until, by November, 100% of males are again encountered only in ones or twos. Females show more variation in group size throughout, and are almost equally likely to be encountered in groups of one to two or three to five throughout in all seasons (Figure 5.4). Larger social groups are, however, more common in April and May; during May the groups break up, females becoming more solitary as they prepare for the birth of their fawns in mid June. By August numbers in herds increase again as does and fawns join with other family groups, and rise once more during September and October as females collect at the rutting stands. Parfitt's data for *mean* group size of such small deer highlight these same trends.

Despite the fact that it may associate into groups of up to 100 or more individuals, the fallow deer is apparently not the truly social species that has so often been claimed. Surprisingly, the basic unit is very much the

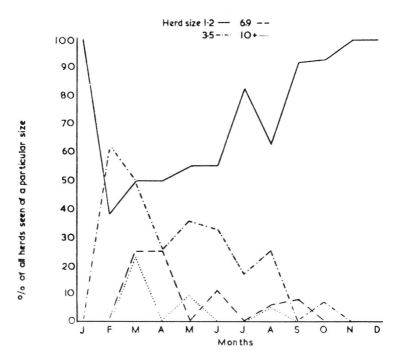

Figure 5.3: Seasonal fluctuations in the size of male deer herds of New Forest fallow deer

Source: From Jackson (1974), with permission.

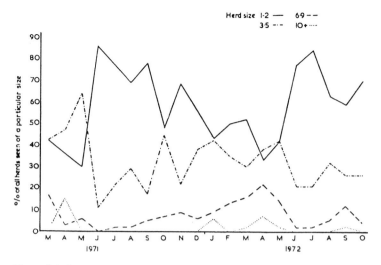

Figure 5.4: Seasonal fluctuations in the size of small-deer herds of New Forest fallow deer

Source: From Jackson (1974), with permission.

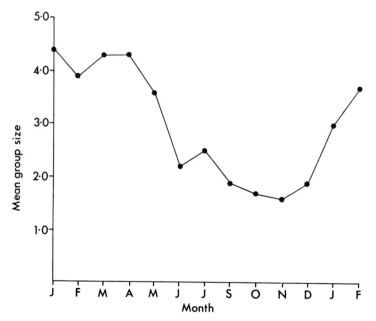

Figure 5.5: Mean group sizes of small-deer-herds observed by Parfitt (1985)

individual; Jackson and Parfitt both make it clear that groups larger than five are infrequent, and it is further clear that even these groups are formed only temporarily and are extremely fluid in composition.

The idea that fallow deer were essentially individualistic was first tentatively suggested by Putman (1981), an idea which was freely admitted to be based on few data and founded largely on hunch. At the time I noted that it was important to distinguish between strict, fundamental social groupings — what might properly be referred to as the social unit — and casual associations of more than one of these units to form larger groups. Thus, while herds of fallow deer of up to 70 or 100 animals may occasionally be observed on favoured feeding grounds in the New Forest, closer observation reveals that this is in fact a coincidental aggregation of a number of separate smaller social units, occurring together merely because they happen to be using the same area simultaneously. If such a 'herd' is observed for a long period, it may be seen that it is not of constant constitution: small groups may be seen to join the aggregation, others to leave. In effect, we must recognise that there are two levels of social organisation; that the feeding groups are no more than casual associations of a variety of sub-units is clear from the fact that, as the group disperses from the feeding grounds, it fragments once more into groups of up to seven or eight individuals, moving off in separate directions. Further, while the smaller social groups are distinct male or female parties, such feeding aggregations may gather both male and female units into a mixed-sex 'herd'.

These smaller groups, of between three and seven or (exceptionally) up to 14 animals, might then be considered a more fundamental social unit. They are indeed the same sort of size of group in which fallow deer are more regularly encountered in other habitats (Figures 5.4 and 5.5). It would appear, however, that these associations are little more stable than the larger aggregations, that even these groups are not persistent, nor constant in composition from day to day. In effect, these too represent only temporary groupings of *individuals*, adopting for the time being a group size adapted to the habitat in which they occur at that time: just as the larger herds are temporary associations of these groups in more open habitats. It may be shown that group size in fallow deer *is* influenced by habitat (Putman 1980), and it may well be that *all* these groupings are formed towards establishment of the optimum group size for exploitation of a particular environment (Putman 1981).[1]

---

[1] The influence of habitat type upon group size has now been studied in more detail by Parfitt, who showed this effect to be relatively minor; in his analyses, no clear differ-

Sika Deer

Sika deer (*Cervus nippon*) are native to eastern Asia, and are found in the wild state in China, the southeastern corner of Russia, Taiwan (Formosa) and various Japanese islands. Populations tend to be of very local geographic distribution — many are island forms, characteristic only of their own tiny island — and this isolated development of different populations has led to the evolution of a number of distinct races or subspecies each characteristic of a small local area. Thus Formosan sika, Manchusan sika and Japanese sika may all be recognised as distinct forms; other races have been described from the mainland, and are still being reported (Zhuopu *et al.* 1978). All races interbreed freely, however, and also interbreed with the more westerly red deer (*Cervus elaphus*). Such ready hybridisation has led to the suggestion that all the various races of sika deer, together with the European red deer and the North American wapiti (*Cervus canadensis*), form a continuous cline or ring species (Harrington 1981); it also suggests that many, if not all, of the mainland subspecies of sika may be entirely of hybrid origin (Lowe and Gardiner 1975).

Sika were introduced to Britain in the 1860s, and to the New Forest in the early 1900s. They have found both climate and vegetation much to their taste and have proved tremendously successful, spreading rapidly through Scotland, Eire and southern England — wherever they have been introduced. So rapidly have they spread that they have reached pest proportions through much of their range, causing significant economic damage to commercial forestry interests in Scotland and Ireland. Their geographical range now extensively overlaps that of the native red deer: the readiness with which the two species hybridise threatens the integrity of the red deer as a distinct species. For both these reasons sika deer in Britain are receiving considerable interest, and much recent research effort has been devoted to this species, for surprisingly little is known about their ecology and behaviour even in their native Asia.

Sika deer are generally regarded as characteristic of somewhat acid soils. 'Typical' sika habitat in Britain could be considered a mix of heathland and coniferous forest (two vegetation types often associated

---

ences in group sizes of small-deer could be detected between different habitats, and the dominating influence upon changing group size remained the seasonal effects of the annual cycle (page 96).

together anyway in that many commercial plantations are established on areas of native heath). The animal is, however, clearly an opportunist and can readily adapt to a wide range of conditions, and diets. Social organisation, grouping, and reproductive behaviour have in the past been little studied, but merely presumed analagous to systems established for the congeneric red deer. Most of what we *do* now know about sika deer ecology has come from two major studies in the South of England: in the Isle of Purbeck and in the New Forest itself (Horwood and Masters 1970; Mann 1983).

Habitat Use
In the 'typical' British habitat of acid coniferous woodland, sika show a very predictable pattern of habitat use. Relatively little forage of high quality is available to the deer within the plantations: when feeding, the deer tend to leave the woodland cover for the open vegetation beyond the forest — in heathland or on agricultural fields. Use of such open habitats is primarily at night (the deer are very sensitive to human disturbance); by day the animals retire to the woodlands and lie up in the dense cover. This same shyness of any disturbance causes the deer to seek out the densest thickets for shelter.

Coniferous forests are not endless, monotonous seas of evergreens that one sometimes imagines. In commercial forests such as Wareham and Purbeck, the varied species composition of different blocks provides diversity of habitat, while the different growth stages of different plots within any forest block results in a mosaic of different structural types. Although all is coniferous forest, one can recognise distinct sub-habitats such as young plantations, prethicket, thicket and polestage forest blocks, as well as mature 'high forest'. The names are self-descriptive, i.e. young *plantations* contain small trees less than 1.5m in height, planted in open ground with extensive and varied ground cover containing several species of grass and many forbs. As the trees grow, they spread and begin to shade out the ground below: a distinct *prethicket* stage may be recognised of trees about 4.5 m in height; the lower branches meet, but do not interlock. These areas provide good shelter, and a certain amount of food: the ground is still grass-covered, and heather, gorse and bramble have often become established. As the canopy closes, and lower branches of adjacent trees interlock, the forest forms a dense impenetrable *thicket.* Ground cover is almost non-existent owing to poor light penetration. At this stage the commercial forest is thinned. Lower branches are brashed, and many trees are also removed, to open up the area and provide room for the final growth of the remaining trees. Such management changes the character of these *polestage* areas: light again penetrates the plantation, to permit some ground cover of grass and bracken; in the thinned areas, an understorey of birch and holly may develop. The conifers continue to grow. Eventually the canopy closes again high above the forest floor, as the trees reach full maturity; this stage is generally referred to as *high forest.*

Despite this variety of habitat types, Horwood and Masters (1970) and Mann (1983) found that sika in Purbeck restricted their activity primarily to one woodland type: the impenetrable thicket-stage plantations. Pattern of use of the range was simple, with the animals lying up in dense thicket throughout the day and moving out onto the heaths or onto agricultural land to feed at night. This regular daily migration was most marked and was similar throughout the year: indeed, the overall pattern of use of the available habitats changed little between seasons (Table 5.3).

In contrast to the coniferous plantations of sika habitat in Purbeck, and through most of their range in Scotland and Ireland, the New Forest offers the deer a much more varied environment — and one of predominantly deciduous woodlands. Sika in the New Forest are (curiously) restricted to a small area in the south of the Forest near Brockenhurst (Figure 5.6), an area seemingly bounded to the north by the Southampton to Bournemouth railway line, to the east and west by the waters of Beaulieu River and the Lymington River. The core of this area is wooded: mostly mature oak-beech woodland with a secondary canopy of holly, yew, hawthorn and blackthorn. There are extensive patches of birch scrub. Within the area, various blocks have been cleared and planted with conifers; all the various structural stages

Table 5.3: Use of available habitat by sika deer in Purbeck Forest (Mann 1983). Numbers are those seen in each habitat as percentage of total sika observed each month during 1981

Jan	Feb	Mar/ Apr	May	June	July	Aug	Sept	
4	4	2	0	0	0	0	7	oakwoods
46	23	40	58	53	59	41	36	thicket
3	4	3	2	6	10	10	7	saltmarsh
10	34	13	7	9	3	8	10	heathland
33	32	41	32	27	24	37	38	fields
4	3	2	2	5	4	4	2	woodland rides

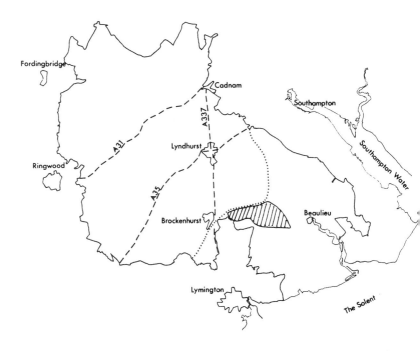

Figure 5.6: The geographical range of sika deer within the New Forest

described above for Purbeck and Wareham Forests are represented. There are many clearings and rides, and the woodlands are surrounded by extensive tracts of heath and a small amount of agricultural land.

With this more varied habitat on offer, sika deer within the New Forest show a more variable pattern of habitat use and show more pronounced seasonal variation in communities exploited. Patterns of habitat occupancy were derived from direct observation and from studies of the accumulation of dung on cleared plots established in each main vegetation type (Mann 1983). Direct observations were carried out using fixed transects, walked or driven on a regular schedule, as in studies of ponies and cattle (Chapter 3). A seasonally changing pattern of habitat occupancy was seen, with most of the animals occupying the oak-beech woodland during the entire winter period and, in spring and summer, continuing to exploit these mixed woodlands after dark but during daylight hours making extensive use of a variety of other habitats, in particular prethicket areas (Table 5.4).

Table 5.4: Habitat use by New Forest sika (Mann 1983). Numbers are those seen in each habitat as percentage of total numbers of sika observed each month during 1981-2

	Jan	Feb	Mar	Apr	May	June	July	Aug	Sept	Oct	Nov	Dec
RV												
SL				not available in environment								
AG/WH	0	0	0	0	0	0	0	0	0	0	0	0
DH	0	0	0	0	0	0	0	0	1	1	0	0
Bog	0	0	0	0	0	0	0	0	0	0	0	0
DW	50	52	50	55	48	31	27	32	35	55	74	62
CW mature	0	0	0	0	0	0	0	0	0	0	0	0
CW thicket and pole	20	17	18	16	16	23	20	20	12	18	10	11
CW clearfell and plantation	6	4	11	16	18	17	23	21	26	7	3	6
WG + rides	24	27	26	13	23	29	30	27	26	18	13	21
Gorse	0	0	0	0	0	0	0	0	0	0	0	0

Animals in Wareham and Purbeck Forests are strongly nocturnal in habit: a pattern of activity no doubt imposed upon them in large part by the fact that they must leave the woodlands to forage in the open, and are less subject to disturbance if using these exposed habitats at night. Sika deer within the New Forest are less subject to disturbance. They may find adequate feed within the woodlands themselves through much of the year, and thus may be found active throughout the 24-hour period in most seasons.

During the summer, the majority of the sika deer in the New Forest feed during the day in small groups in prethicket areas (although all habitats are used to a certain extent), exploiting the extensive forage supply and benefiting from the security of the relatively dense habitat. Diet at this time of year shows a high intake of grass and leaves (see Chapter 6), and these items are readily available in all habitats. At night, still in small groups, most of the animals lie up in the Forest oakwoods or in the shelter of the extensive polestage areas. A few deer may be found ruminating in the open.

In the autumn, when acorns and leaves fall, the characteristics of a large part of the New Forest habitat change rapidly. The prethicket still provides a good food supply, but is now inferior in quality to that offered by the oakwoods, which also provide a considerable amount of shelter. Unlike the rest of the year, when it could be suggested that the need to reduce intraspecific competition precludes the formation of large feeding aggregations in this habitat, the relative abundance of acorns and leaves now permits larger feeding groups to collect. This period, however, also coincides with the rutting season and it has been suggested that the males may also provide a focus for the accumulation of groups of small-deer at this time. In late autumn and in winter, when the acorn crop has been exhausted and most of the fallen leaves are decaying, food supplies within the Forest are more limited and the deer start feeding more extensively on coniferous browse and *Calluna* (Chapter 6). These are available in a variety of habitats; the animals still spend most of their time within the oakwoods, but may forage out in prethicket and heathland areas.

Social Organisation

In an analysis of over 3 observations, Mann (1983) found a clear annual cycle of group size in New Forest sika (Figure 5.7), similar to that discussed earlier for fallow. Once again the basic social 'unit' is the individual (male, female, or female with calf), but in sika this is far more obvious; indeed, sika are one of the less social of the deer species and

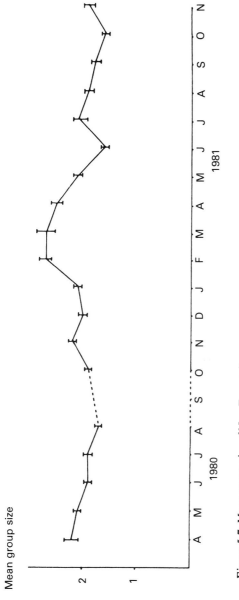

Figure 5.7. Mean group size of New Forest sika

Source: Mann (1983).

through much of the year the individuals remain completely solitary. From the end of winter through until September, animals are generally encountered alone, or as hind and calf. The rut in September causes an increase in group size, and increases the number of groups encountered of mixed sex; rich autumn food supplies allow these larger aggregations to persist through until March or April. Even at this time, however, sika never establish groups anything like as extensive as the feeding herds of 100 or more strong which may be formed by fallow. Indeed, they are rarely observed in groups of more than five or six; the largest group ever recorded in Europe consisted only of 12 individuals.

These larger feeding groups are temporary aggregations only and show little constancy of composition (though individual hinds associating with the same stag through the rut may be seen together on a more regular basis). Horwood and Masters (1970), working in Purbeck Forest, succeeded in individually marking a number of sika hinds with colour-coded collars. Successive observations showed a fluidity of group composition much as reported on page 100 for fallow. Although each hind had a relatively small home range and might be seen feeding in the same fields or same group of fields on a number of successive occasions, her companions differed. Horwood and Masters quote the example of a hind with a blue and yellow collar seen on 23 August 1969 in the company of four other hinds and five calves, one of the hinds having a red and yellow collar; two weeks later, she was seen in a different field 600 m away with three unmarked hinds and four calves. The pattern was repeated many times with different marked deer (Horwood and Masters 1970). Such observations confirm the conclusion that, at least in Purbeck, the individuals are independent members of a very loosely knit community which, since ranges show extensive overlap, may be seen feeding together in the same areas at night, but lie up independently during the day. Sika in the New Forest give the impression of similar independence.

Despite this apparent independence, all the animals within a particular population are clearly aware of each other; flexible group composition means that each individual will regularly come into contact with all other individuals sharing its range. The individuals in a particular area thus form a loose 'clan' or 'superherd' (*sensu* Putman 1981), and it seems probable that this group has a real social meaning and is more than just a passive 'envelope' of independent individuals defined together merely because of geographical coincidence. Putman (1981) noted:

Since the animals of a given area of forest will tend to use the same feeding grounds and their total range is relatively small, it is of course inevitable that one is likely to see a relatively small set of animals together fairly frequently and that they will not interact with a set of social units whose ranges are polarised on a totally different set of feeding grounds. The "clan" suggested may thus be no more than a statistical artefact. Nonetheless, it is very tempting to postulate within any one area a form of "superherd" or "clan" — an extended social network of independent but related social groups which may merge and separate and interact, but which never interact with any group outside the "clan": a system analogous to that of lion "prides". Such a supposition sweeps away other apparent inconsistencies between "herding" and "non-herding" species of deer — and may also help to explain the operation of the social units of the "herding" deer in their apparently random association into different feeding groups.

At least among sika, it would appear that each geographical population, however loosely knit, does have some social ties. Mann (1983) showed for New Forest sika evidence highly suggestive of a dominance hierarchy among hinds of overlapping home range. Direct interactions between individuals were observed too infrequently (and most hinds are indistinguishable, anyway) to allow the construction of a dominance hierarchy. That one exists, however, was suggested by the fact that some feeding hinds may be displaced by others and that in a feeding aggregation the deer seem to be relatively evenly spaced. Mann interprets such observations as suggesting that a dominance hierarchy has been established earlier and is maintained by signs that are, on the whole, too subtle for the observer to detect.

This is in spite of the fact that a high proportion of the hind's life is spent in habitats where small groups are maintained and where the opportunities to establish such a hierarchy must be limited. Mann (1983) suggests that a temporary hierarchy may be established in these feeding aggregations of sika deer, or, that if the same animals feed together on subsequent occasions as members of the same superherd or clan, then the hierarchy may be gradually established over a longer period and merely consolidated on these occasions.

Results of Horwood and Masters's work in Purbeck, and Mann's later studies in Purbeck and in the New Forest, accord well with other studies of social structure of sika in the wild. Group sizes are universally reported to be small; in Japan, the mean group size recorded is less than three, and most groups contain a hind and calf or a single male (Miura 1974). Larger groups, containing between five and 11 individuals, are sometimes formed during the winter months. In Russia, Prisyazhnyuk and Prisyazhnyuk (1974) report that the most frequently encountered groups number from two to ten animals, and in Poland, where sika deer were also introduced at the turn of the century, Dzieciolowski (1979) reports groups most commonly being between two and six animals, with the largest recorded numbering 12 individuals. All studies report largest aggregations in autumn and winter, with animals relatively solitary for most of the year; herds of mixed sex are formed during the rut and persist over winter, contributing to the increase in group size.

Part of the explanation for the observed increase in group size over autumn and winter in the New Forest is a change in food availability and habitat use. The rich mast fall provides an abundant, but highly localised food supply. It is a well-established phenomenon that not all areas of woodland, not all individual trees in any one area, crop well in any one year. While there may be generally 'good' mast years and years of poor fruit fall, in general different areas of a woodland are not at all well synchronised and mast fall is extremely patchy. As a result, this rich food source is very local in distribution; to exploit it, the animals must aggregate in relatively small areas. At the same time, in areas which have fruited, the mast is generally abundant and thus, with relatively little competition for food, larger groups can be tolerated.

Although this is a special case, habitat occupied does have an effect on group size and composition throughout the year. At any one time of year, group sizes on open habitats are larger than those of closed vegetation such as oakwood or prethicket. The difference is not particularly large (as might be expected when group sizes range only from one to, exceptionally, six individuals), but is consistent with results reported for other ungulate species (e.g. Estes 1967, Jarman 1974; and see above, page 100) where an inverse relationship may be demonstrated between group size and cover density. Relationship of group size to habitat type is precise: in habitats which are represented in both Purbeck Forest and the New Forest (oakwoods, rides, fields and prethicket), group size was the same in the two areas (Mann 1983).

As is the case with most deer species, sika hinds and stags lead very separate lives for most of the year. Although mixed-sex groups of sika may be encountered in almost any month, these are primarily chance feeding aggregations (cf. mixed-sex herds of fallow deer, p. 100) and they are most frequent around the time of the rut. Once the rut is over, mixed groups of stags and hinds may still be seen feeding together quite commonly for a month or two, but gradually the two sexes separate into their own distinct groups and remove to their own distinct ranges.

During the rut, males return once more into the main population to breed. Horwood and Masters (1970) report sika rut in Purbeck as taking place between early September and the end of October; in the New Forest the rut appears to start somewhat later, not reaching its peak until mid October. The first indication of sexual activity is the increase in calling by the males. This extraordinary noise — a shrill whistle repeated two or three time — may be heard occasionally throughout the year, but becomes more frequent during late September and October.

Published information on the rutting behaviour of sika deer (Horwood and Masters 1970, for Purbeck) suggests that the stags mark out and defend territories in the woods: territories which the hinds have to cross in their nightly migration to the fields to feed. These territories are marked by thrashed *Calluna* bushes and frayed perimeter trees, and Horwood and Masters report that the stags remain on their territories throughout the main part of the rutting season and do not resume feeding in the fields until the end of October.

At this time of year the master stag tolerates the presence of younger, attendant, stags, who may accompany him much in the same way as junior stags accompany the harem master in the red deer (Clutton-Brock and Albon 1978). Fighting between stags, Horwood and Masters report, is commonplace, both between the mature male and one of his attendants and between neighbouring territory holders.

From Mann's observations in the New Forest, however, it is not clear whether sika stags in this area mark out and defend a territory in this way or whether they collect and defend a harem as do red deer (Lincoln *et al.* 1970; Lincoln and Guinness 1973). Certainly, stags are found associating continuously with groups of females. When disturbed, the male is the last to depart and on a few occasions his actions could be described as 'rounding up' the hinds. On the other hand, males are frequently encountered alone: so either the area contains many unsuccessful males and a few dominant individuals who collect the harems, or the stags 'float' and cover receptive females as they find them.

The few animals which could be individually identified seem to spend the rutting season in a relatively small area of the woods, and return to the same area year after year, but these areas do not appear to be established as exclusive territories, as suggested by Horwood and Masters for the Isle of Purbeck. Where ranges could be plotted at this time, a considerable degree of overlap is observed (Mann 1983) and there is no evidence of boundary marking by thrashing *Calluna* bushes or fraying of perimeter trees. Some tree damage is observed (most commonly scoring of the bole or trunk with the antlers) and individual trees may be marked in several successive years; the damage is, however, unrelated to location (Carter 1981 found that in some cases over 90% of the trees in one compartment were damaged in this way) and seems unconnected to territory marking.

Overlaps in males' ranges and lack of boundary marking in the New Forest, coupled with relatively low calling rates and few observed fights, led Mann to conclude that in the New Forest sika stags do not defend a territory. Yet nor do they appear strongly to select hinds and hold them in a fixed harem. Mann concludes that the stags seem to patrol areas of superior food quality, and cover hinds in oestrus when they find them: perhaps a more appropriate strategy within the New Forest, where animal movements are more 'internal'. Territoriality at Purbeck may be linked to the regular daily migration of hinds from cover out into open fields at night to feed: territories are certainly concentrated at the woodland edge. In the New Forest, where such regular daily movements do not occur, a different strategy appears more favoured.

Red Deer and Roe

Red deer and roe are now uncommon in the New Forest, and little work has been undertaken as yet on their general ecology.

Red deer have always maintained themselves as essentially local populations — and, curiously, populations have been continuously 'subsidised' by introductions. Both James I and Charles II introduced fresh blood from France, Charles II importing no fewer than 375 deer which were released near Brockenhurst; further introductions continued throughout the nineteenth century and even into the early twentieth century.

Census records are patchy: during Henry VII's reign there were several records of red deer being killed in the New Forest, and in 1670 numbers were estimated at 357 head: 103 male and 254 female deer. By the late eighteenth century, however, the Forest's red deer population was probably extinct; certainly, returns of 1828-30 on deer in Royal Forests omit any mention of red deer within the New Forest. References to sightings in the nineteenth century probably relate to park escapes (the population of 15-20 head reported in 1892 was possibly an isolated party which had wandered across from the Wiltshire border), and in the last 200 years numbers have probably never exceeded 80-100 head.

Present-day populations are focused in two main areas of the Forest: in the northeast around Burley, and in the southeast around Brockenhurst. Southerly populations are said to be descended from three animals released by Lord Montague in 1908 into Hartford Wood. The more northern population may possibly date from much the same period. At the beginning of the present century it was not unusual to see herds of red deer in Milkham, Slufters and Holly Hatch Inclosures near Burley — in much the area still occupied to the present day. This population, too, has, however, suffered introductions, when in 1962 a number of animals escaped from Sir Dudley Forwood's estate in Burley.

Roe may be presumed to have been native to the New Forest area, but during the Middle Ages the species became virtually extinct throughout England, with the exception of the Border Counties and a few isolated sites in the South. As with red deer, the modern populations of roe in southern England have resulted from the reintroduction of animals into several places during the nineteenth and twentieth centuries (Prior 1973). Roe recolonised the New Forest from 1870 onwards, spreading across from Dorset (Jackson 1980). Census figures suggested a population of perhaps 400-500 animals in the early 1970s, but since that time populations have steadily declined. Numbers are now extremely low (estimated by the Forestry Commission as 264 in 1984), and breeding performance of those animals which do remain is noticeably poorer than average for the species: New Forest roe does never conceive before their second year and usually have only a single offspring (the national average is 1.8 kids per female: Rowe, pers comm.). The decline is unexplained, although it is generally believed that it is due to changes in habitat over the past 20 years.

Despite the fact that the roe is an opportunistic species — and may be found in a wide variety of habitats from dense woodland to open agricultural land with little or no cover (e.g. Zejda 1978, Turner 1979, Kaluzinski 1982) — it is primarily a pioneer species, associated particularly with early successional communities. Such communities are characterised by rapid growth and high production of the 'concentrated' foods that the roe favour; hence roe are themselves at their most productive in young woodlands or other disturbed habitat which offer a rich ground vegetation. Young deciduous woodlands where light can still reach the ground level offer an abundance of grasses and forbs, shrubby vegetation such as roses and brambles, and young understorey saplings of hawthorn, hazel or willow. Farmland offers rich pasture, the nutritious tips of growing cereals or an abundance of dicotyledonous weed species among the crops. Among coniferous plantations, too, younger plantings will support more productive roe populations. The wide spacing of the young trees allows development of grass and forbs in the gaps, once again with other ground-layer shrubs such as rose and bramble; the trees, too, are at a height where roe can, if they wish, browse on the growing buds (although roe do not particularly favour coniferous browse: Chapter 6). As the plantation matures, the canopy closes and cuts out the light reaching the forest floor: the ground flora is eliminated and the trees themselves are too tall to offer food. Roe *are* still found in such plantations, but use them almost exclusively for cover, foraging out from such shelter into younger woodland blocks or open vegetation at dawn and dusk. As a result, such plantations can support good populations of roe only if they are associated with open ground nearby, or other less advanced blocks of woodland. In most commercial softwood forests, trees are planted in blocks and on a rotation: so that at any one time a single forest will contain blocks of trees of all ages, from areas newly planted to stands of mature timber. The New Forest has, however, had a rather curious management history. While considerable planting was once carried out over extensive areas, there has been relatively little planting in more recent years. The earlier stock is now pre-thicket or thicket (page 102) offering little fodder, yet alternative foraging areas are rather sparse.

Roe deer are now extremely scarce in the New Forest, and indeed occur in any density only in two small areas (where open grazing is still available: around Ashurst and Godshill), and it seems highly likely that changes in habitat may be largely responsible for this decline. The change in vegetational structure, however, is not restricted just to the effects of ageing on the Forest woodlands. We have already noted that a primary influence on Forest vegetation is that of the grazers themselves, and it is possible that the decline in roe populations is also in part due to competition, direct or indirect, with other herbivores. Roe populations, as might be expected, have always been associated with the Forest woodlands. Most of these are enclosed and, at least until the 1950s, relatively free from grazing by commonable stock. More recently, however — and at about the same time as roe populations began to decline — the exclusion of, particularly, ponies from Statutory Inclosures has been less rigorously enforced and ponies are now quite commonly encountered in most Inclosures. The resultant effect on the vegetation is striking. Rides and ride edges (which used to have to be cut two or three times a year for rabbit control: Courtier, pers. comm.) are now close-cropped all year. Available browse is removed to a height of up to 2 m — and the height to which a determined Forest pony can stretch is greater than that of a roe deer. Perhap this recent influx of Common animals into Inclosures, and *their* effect on the availability of suitable forage, has also contributed to the decline of roe. Direct dietary overlap between the two species, and between roe and the other main herbivores, is considered in more detail in Chapter 6.

Chapter 6
Food and Feeding Behaviour of the Forest Deer

Much detailed research has been undertaken on the feeding behaviour and diet of deer in the New Forest (e.g. Jackson 1974, 1977, 1980; Mann 1983; Parfitt 1985) and consequently we know a great deal about the food habits of the major species: fallow, roe and sika. Patterns of habitat use in foraging are virtually identical to those described for overall patterns of habitat occupancy: foraging largely dictates habitat choice, and changes in food availability are primarily responsible for seasonal change in habitat use. As a result, we may concentrate in this chapter on the actual diet and dietary intake of the three species: comparing dietary composition in the New Forest with that recorded for the deer elsewhere in their range, and considering the degree of dietary overlap experienced by the three species where they co-occur in the New Forest — a potential source of competition in seasons of food scarcity.

Fallow Deer

The diet of fallow deer within the New Forest has been studied in detail by Jackson (1974, 1977) and more recently by Parfitt (1985). Jackson's results are based on identification of macroscopic plant remains in rumina of culled animals; Parfitt's reconstruction of dietary composition derives in the main from microscopic analyses of faecal material. This is not the place to embark on a discussion of the merits and demerits of ruminal or faecal analyses as methods of diet determination and the interested reader is referred to, for example, Staines (1976) and Putman (1984).

John Jackson's analysis of the diet of fallow deer was impressively thorough: his results came from painstaking examination of rumen samples from 325 deer over the period November 1970 to March 1973.

His results (Table 6.1) show that through most of the year the deer are primarily grazers. Throughout the growing season — from March to September — grasses form the principal food (comprising in the region of 60% of total food intake), with herbs and broadleaf browse also making a significant contribution. Acorns and mast are a characteristic food throughout autumn and early winter, although their importance in the diet varies from year to year with variations in the mast crop. Other major foods through autumn and winter, and on which the deer rely more heavily when the year's mast is exhausted, are bramble, holly, ivy, heather, and browse from felled conifers. Even at this time, however, grass still makes up more than 20% of the diet. It is evident from this that the deer are preferential grazers throughout the year, and take increasing amounts of browse over autumn and winter merely to compensate for lack of graze materials outside the growing season.

Jackson's data present an extremely thorough analysis of diet of New Forest fallow. The amounts and proportions of different food items ingested have not, however, been related to the *availability* of those foodstuffs in the environment: we do not know how selective the deer may be in their feeding. Further, since we have no knowledge of how the abundance or *quality* of the various forages may change at different times of year, we cannot fully appreciate either why the animals choose the forages they do or what causes them to change their diet in different seasons. Parfitt's later analyses of the diet of New Forest fallow were, however, undertaken within the context of detailed knowledge of changing availability and quality of forage over the course of the year.

Despite the fact that Parfitt's analyses were based upon different methods (the examination of faecal rather than ruminal materials) and represented the diet over a totally different time period, some ten years after Jackson's study, basic results were strikingly similar to those of the earlier work, suggesting an identical pattern of forage use. With some confidence that the results obtained thus present a realistic picture of fallow diet within the Forest, Parfitt then went on to test the diet selected against a number of measures of availability and nutrient quality of the different forages taken (available standing crop, energy value, and nitrogen, calcium, potassium, phosphorus and magnesium content). The one consistent relationship that emerged was with digestible nitrogen: over the winter months the deer appeared to select those forages highest in available protein. Interestingly enough, this correlation between nitrogen content of forages and their importance in the diet seems to hold only between November and March. Over the sum-

Table 6.1: Percentage contribution of the main forage components to diet of New Forest fallow deer (Jackson 1977)

	Jan	Feb	Mar	Apr	May-July	Aug	Sept	Oct	Nov	Dec
grasses	21	25	59	67	63	57	58	33	25	21
herbs	1	1	1	6	6	12	7	2	2	1
conifers	14	14	7	1	0	0	0	0	8	17
holly	12	17	9	7	4	3	1	0	2	7
other broadleaves	1	1	0	5	14	11	6	4	12	4
heather	16	24	16	3	4	3	2	1	8	16
bramble/rose	17	7	2	3	3	12	7	10	10	8
ivy	8	4	1	1	4	1	2	0	4	4
gorse	0	0	0	0	0	0	0	0	0	0
fruits	2	0	0	0	0	0	14	41	22	15
other	8	7	5	7	2	1	3	9	7	7

mer growing season — when perhaps *all* types of forage offer adequate protein? — the relationship breaks down; it is only during the winter months, when we may assume that there may be greater variability in forage quality, that there appears this positive selection for foodstuffs richest in digestible nitrogen.

Parfitt's data for diet composition are based on faecal analysis. If we know the total faecal output of each fallow deer, we can estimate total daily intake of forage by the deer in much the same way as we calculated total dietary intake for cattle and ponies in Chapter 4. Adult fallow deer on open range are observed to defaecate between ten (Bailey 1979) and 12 (Putman, unpublished) times during a 24-hour period. From theoretical considerations (density of fresh pellet groups in relation to known densities of deer: Bailey and Putman 1981), a suggested mean figure of 11.33 defaecations per day is derived for use here. Mean dry weight of 20 pellet groups from adult deer (males and females) in the New Forest each month has more recently been determined (Putman 1983). If these figures are multiplied by 11.33 (number of defaecations per day), total faecal output per animal per day in each month may be estimated. Such material has, however, suffered loss from digestion and the total defaecated weight does not in itself equate to ingested weight.

Parfitt's estimate of diet is, however, based on proportion of forage fragments *in faecal material*. If these figures are thus multiplied by the assumed daily faecal output, we may derive a figure for total *faecal* output, by weight, of each forage species as on page 79). If these figures are themselves further corrected by forage-specific digestibilities (multiplied in each case, by $100/100$-DMD% — see page 79, and for fuller justification see Putman 1983), we may arrive at an estimate for total *ingested* weight of each forage material (Table 6.2). Summed over a month, such results allow derivation of total dry-matter intake. Results are interestingly of the same order of magnitude as voluntary food intake recorded in feeding trials on captive fallow deer (Putman 1980), thereby offering some confidence in our reconstruction.

Table 6.2 presents daily dry-matter intake of adult fallow deer, broken down by specific forage categories. Chemical composition of each forage species (digestibility, energy content, N, P, K, Ca) in each season is in this case known (Parfitt 1985; and see above, page 78). In any one month, therefore, the estimated ingested weight of each forage species may be multiplied by the laboratory-determined value for content of each nutrient (derived for that particular forage, in that particular area in that month). Summed within any month, over all forage species, these figures offer an estimate of total dietary intake of each of the

Table 6.2: Total ingested weight of different forage classes by New Forest fallow deer from ruminal analayses of Jackson (1977) (Values in g, dry weight): see text for derivation

Month	Broadleaved trees	Mast	Conifers	Ilex	Rubus and Rosa	Hedera	Ulex	Heathers	Other shrubs	Grasses	Ferns	Mosses	Forbs	Others	Total (dry-matter intake)
								Dietary category							
January	3.7	7.3	51.2	43.9	62.2	29.3	0	58.6	7.3	76.8	7.3	3.7	3.7	10.9	365.9
February	3.7	0	51.2	62.2	25.7	14.6	0	87.8	7.3	91.4	3.7	7.3	3.7	7.3	365.9
March	0	0	21.5	27.7	6.1	3.1	0	49.2	12.3	181.3	0	3.1	3.1	0	307.4
April	12.5	0	2.5	17.4	7.5	2.5	0	7.5	10.0	166.8	2.5	2.5	14.9	2.5	249.0
May-July	26.7	0	0	8.2	6.2	8.2	0	8.2	4.1	129.7	0	2.0	12.4	0	205.7
August	28.3	0	0	7.7	30.9	2.6	0	7.7	2.6	146.8	0	0	30.9	0	257.5
September	14.4	33.6	0	2.4	16.8	4.8	0	4.8	0	139.2	0	0	16.8	7.2	240.0
October	8.9	91.3	0	0	22.3	0	2.2	2.2	0	73.5	0	0	4.5	17.9	222.8
November	26.2	48.0	17.4	4.4	21.8	8.7	0	17.4	10.9	54.5	2.2	2.2	4.4	0	218.1
December	8.5	32.1	36.3	14.9	17.1	8.5	0	34.2	4.4	44.8	2.1	0	2.1	8.5	213.1

nutrients studied. Results are presented in Table 6.3. Comparisons of these data with results obtained for nutrient intake of captive deer on diets of different qualities (Putman 1980, 1983) suggest that, for most of the year, New Forest fallow deer attain levels of food intake equivalent to those sufficient to keep captive deer in extremely good condition. *Winter* intake of digestible nitrogen (perhaps the most crucial of all nutrients), however, fell to rather low levels (1.4-1.6 g), barely sufficient to maintain condition. Comparisons of actual and required intake of all nutrients considered, in relation to animal body condition, suggest a lower desirable daily intake for fallow deer of each nutrient as shown in Figure 6.1. (The rationale behind this conclusion is developed in full in Putman 1983.) It appears that New Forest fallow deer attain at least minimum intake of all nutrients in all seasons: a result which emphasises the success of their flexible feeding strategy and seasonally changing dietary composition, in enabling them to maintain adequate

| | Digestible energy | Digestible | | | |
		N	P	K	Ca
Per animal	1,600	1,500	300	2,300	1,200
Per kg body weight$^{(0.75)}$	70.7	65	15	100	50

Figures in mg/day, except energy (kJ/day)

Figure 6.1: Suggested minimum required intake by fallow deer of key nutrients

Table 6.3: Estimated daily nutrient intake of New Forest fallow deer (Dietary profile from Jackson 1977) Figures are presented as intake of each nutrient (in grams) and energy intake (KJ)

	N	digN	P	K	Ca	E	digE
January	6.46	2.64	0.61	3.19	2.78	7,270	2,994
February	5.99	2.38	0.67	2.90	2.73	7,290	2,880
March	5.46	1.77	0.58	2.45	1.54	5,772	1,804
April	5.11	1.66	0.57	2.44	1.30	4,550	1,406
May–July	6.06	2.02	0.35	3.42	1.28	3,680	1,672
August	5.66	2.16	0.40	3.96	2.11	4,806	1,909
September	4.44	1.78	0.38	3.21	1.88	4,520	1,926
October	3.66	1.61	0.32	2.36	0.99	4,378	2,135
November	3.63	1.43	0.27	1.91	1.49	4,380	1,826
December	3.29	1.31	0.29	1.70	1.52	4,326	1,706

nutrition throughout the year despite the changes in quantity and quality of available forage.

Sika Deer

Mann (1983) used both ruminal and faecal analyses in complement, in derivation of a picture of diet for sika deer in southern England. The complete annual diet of both Wareham Forest and New Forest sika was determined through identification of plant cuticular remains in fresh faecal pellets; information on winter diet was supplemented by analysis of rumen samples from culled animals.

At Wareham, the diet was shown to be relatively constant throughout the year, with a high intake of grass (30-40%) and *Calluna* (40-50%) in all seasons; a variety of other dietary components contribute to the remaining part of the diet, but no single item comprised more than about 8% at any time (Figure 6.2). Grasses consumed were principally *Molinia caerulea* (50% of all grass), *Agrostis setacea* and *A. tenuis* (Mann 1983).

This dietary profile — a diet composed primarily of grasses and heather and seasonally unchanging — is not peculiar to Wareham, but is in fact rather generally characteristic of British sika. While investigating the diet of New Forest and Wareham animals, Mann *also* undertook an analysis of diet of Scottish sika deer from a variety of locations for comparison. Rumen samples were collected over two culling seasons (1979-80 and 1980-81) from five commercial coniferous forests in Scotland, where sika deer are becoming increasingly abundant. Results did not differ significantly between forests — nor between years — and results have been pooled to present a profile of general winter diet for Scottish sika deer (Mann 1983). The diet is extremely similar to that described for Wareham sika except that the Scottish animals take less *Calluna*, with grasses comprising 70% or more of their diet.

In general, then, sika deer in Britain may be considered primarily grazers, a result compatible with the limited data available for their diet in their native range. Prisyazhynuk and Prisyazhynuk (1974), writing of sika deer on Askold Island, USSR, report that the bulk of the food is of grasses; Furubayashi and Maruyama (1977) found that sika in the Tanzanawa Mountains in Japan consume 106 plant species but feed primarily as grazers. Takatsuki (1980) used faecal analysis and direct examination of feeding site to investigate diet of sika on Kinkazan

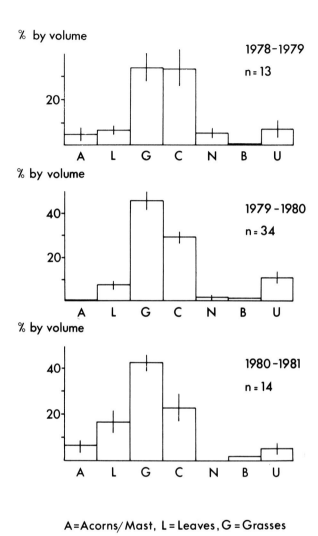

A=Acorns/Mast, L = Leaves, G = Grasses
C =Heather, N = Pine Needles, B = Bark
U = Gorse

Figure 6.2: Winter diet of sika deer in Wareham Forest, Dorset

Source: Mann (1983), with permission.

Island, off Japan, and again concluded that the main dietary component was always grasses (specifically *Zaysia japonica, Miscanthus sinensis,* and *Pleioblastus chino,* a dwarf bamboo which by virtue of being an evergreen is taken all year round).

By comparison, diet of the New Forest animals appears somewhat unusual, for New Forest sika take considerable quantities of deciduous and coniferous browse, particularly in the winter. This curiosity was initially pointed out by Horwood and Masters (1970). Although their data were based on examination of 200 rumina from the Isle of Purbeck but only six New Forest rumina, Mann's later analysis (1983) confirms that New Forest animals rely heavily on browse over winter, when it may comprise up to 23% of total dietary intake. In addition, the animals show striking seasonality in diet, feeding very opportunistically on foods as they become available (Table 6.4).

In spring and summer the New Forest sika do feed extensively on grasses (about 30% during spring, up to 40% in summer) and *Calluna* (30% in spring, 35% in summer), as do Purbeck and Scottish populations. But the diet is far more varied, and includes significant amounts of other forages, forbs, deciduous-tree leaves, gorse and conifer needles. In autumn only 25% of the diet is formed of heather and grass, the remainder being composed of pine needles (from coniferous browse), gorse, holly and acorns. Winter sees an increase in intake of needles and, as might be expected, a decline in intake of deciduous browse and forbs; at this time, less than 20% of the diet is made up of grasses.

Such a heavy emphasis on browse throughout the year is unusual, and cannot easily be explained in terms of changing forage availability or quality. Analyses were undertaken to see if there is any correlation between the proportion of any item in the diet in a particular season and the area of habitat likely to contain that feedstuff (Mann 1983), and in most seasons there is good correlation. It is clear, however, that the deer are still preferentially grazers — and in the summer, when grass is more abundant, correlation fails because few coniferous needles are taken despite their abundance. Further, dietary intake does *not* correlate with nutrient status in the same way that was demonstrated for fallow deer or for New Forest ponies (page 79). Mann concludes that in the New Forest sika deer may be forced by competition with other herbivores to take large quantities of browse material. Although roe and fallow deer are uncommon in that part of the Forest where the sika occur, horses are abundant. As we have noted, New Forest ponies are officially excluded from the Forestry Inclosures, but in practice they are free-ranging and consume so much of the ground vegetation that the practice of cutting

Table 6.4: Percentage contribution of different foodstuffs to diet of New Forest sika deer (Mann 1983)

	Jan	Feb	Mar	Apr	May	June	July	Aug	Sept	Oct	Nov	Dec
grasses	25	25	22	39	38	40	39	50	44	31	28	27
herbs	0	0	0	0	0	0	0	0	0	0	0	0
conifers	20	19	23	13	2	0	0	1	0	6	8	16
holly	1	3	1	1	1	1	1	1	2	2	1	1
other broadleaves	11	11	10	13	14	14	16	10	19	25	14	14
heather	24	23	30	23	35	37	35	29	27	23	24	25
bramble/rose	0	0	0	0	0	0	0	0	0	0	0	0
ivy	0	0	0	0	0	0	0	0	0	0	0	0
gorse	7	14	8	7	6	5	6	6	7	4	7	5
fruits	6	4	2	0	0	0	0	0	0	6	14	9
other	6	1	4	4	4	3	3	3	1	3	4	3

the grass on the rides to reduce the fire risk (common until a few years ago) is no longer necessary. In the summer, forage productivity is sufficient to support offtake; in the winter, however, food resources become more limited. The ponies, by virtue of their opposed incisor teeth, can graze closer to the ground than the deer and, as their intestine relies upon a rapid throughput of large volumes of material, it might be expected that large amounts of grass will be rapidly consumed. It seems possible that the deer cannot compete effectively at this season and this may be why they feed on coniferous browse, a resource not exploited by sika in other areas (Mann 1983).

Roe Deer

Although, owing to their low density within the Forest, little work has been done on the behaviour and general ecology of roe deer in the New Forest, their diet has been well researched. While carrying out his classic studies of fallow deer (Jackson 1974), when he was collecting regular samples from the Forestry Commission cull for analysis of dietary composition, John Jackson took the opportunity to collect in addition samples from any roe deer which were shot, killed in road-traffic accidents or otherwise died within the Forest: 105 rumen samples were collected between November 1970 and March 1973. Jackson's results (Table 6.5) showed that young browse material (bramble, rose, and new growth of deciduous woodland trees and dwarf shrubs) formed the bulk of the diet through the year. Bramble and rose together formed between 25% and 45% of the diet throughout, and comprised the largest food fraction in all months except January and April. From January to March, foliage from felled conifers or from young plantations, *Calluna* and ivy were major foods, with lesser amounts of grasses, herbs and fungi. Over the summer, growing season, herbs and grasses became more important, and new growth of deciduous trees and shrubs was also favoured. (Although browse from these plants was seen in the ingesta all through the year, amounts present from October till March were minimal; from April to September, however, such browse formed between 10% and 30% of the diet.) During the autumn, acorns were a characteristic food, when available; fruit and nuts formed up to 15% of the diet from September through till midwinter.

Browse species used through the summer included birch, beech, oak, hawthorn, willow and buckthorn, as well as holly and shoots of bilberry. Twigs and foliage of coniferous species (primarily Scots pine and Nor-

Table 6.5: Percentage contribution of different foodstuffs to the diet of New Forest roe deer (data from Jackson 1980)

	Jan	Feb	Mar	Apr	May	June	July	Aug	Sept	Oct	Nov	Dec
grasses	4	5	5	10	7	8	8	8	8	9	10	4
herbs	5	2	2	30	16	16	16	17	17	4	4	6
conifers	33	22	22	5	1	8	0	0	0	12	12	13
holly	0	2	0	0	0	14	1	0	0	0	0	0
other broadleaves	2	4	4	14	30	14	15	15	14	5	5	2
heather	6	14	14	14	5	4	7	7	5	5	4	7
bramble/rose	31	26	26	20	35	32	40	40	38	38	37	46
ivy	12	22	22	7	6	4	3	3	3	11	11	2
gorse	0	0	0	0	0	0	0	0	0	0	0	0
fruits	1	1	1	0	0	0	0	0	8	8	17	7
other	6	2	3	0	0	0	10	10	7	8	0	13

way spruce) were taken throughout the year, but were most important from October to March. Grasses taken included *Agrostis tenuis, A. canina, Deschampsia flexuosa, Poa annua, Holcus lanatus* and *Sieglingia decumbens.* Roe deer seemed to avoid *Agrostis setacea, Molinia caerulea, Deschampsia cespitosa* and *Brachypodium sylvaticum* (making interesting comparison with New Forest ponies, page 73, and sika deer, page 123).

Roe have long been known as highly selective feeders who pluck small and highly nutritious morsels from a wide variety of plant species. Although it is often assumed that such a feeding style necessarily equates with browsing habit, such an assumption is unfounded: close examination of all published data shows that roe have always been known to take considerable quantities of grasses and forbs when these are at their most nutritious. Their diet in the New Forest, as described by Jackson, is very similar in general 'shape' to the diet found by other workers in Great Britain and elsewhere in Europe. Because food availability and precise species composition varies so much between different localities, detailed comparisons of actual species eaten or precise composition of diets from different areas are of limited value. Perhaps the most useful comparisons which may be made here, just in illustration, are between the diet of New Forest roe deer as described by Jackson (1980) and that of roe deer in other primarily deciduous woodland areas elsewhere in the South of England (Hosey 1974, Johnson 1984). Hosey's data for Dorset show that, as in the New Forest, bramble was the main food in all seasons; broadleaved deciduous browse was important in the summer, and again use of herbs and grasses peaked in the spring and summer growing seasons. In Hosey's study, herbs formed one-fifth of total intake in March and about 15% in November and December; consumption of grass peaked in May and November. Intake of coniferous browse was restricted primarily to the winter, as in the New Forest, but reached its maximum value in November at only 9% of total intake. Perhaps the most notable difference in diet between New Forest roe and those elsewhere in Britain is that ivy figured substantially in the winter diet only in the New Forest (Jackson 1980).

Dietary Overlap

Diets have been described in detail for the three major deer species of the New Forest; and it is clear that they are markedly different. Fallow deer are primarily grazing animals: from March to September grasses and forbs form the principal food; during autumn and winter this diet is supplemented with acorns and beech mast when available, and increased intake of browse (coniferous browse from felled conifers, bramble, holly, *Calluna*), but even over the winter grasses are still the major food. Roe have been shown to have a mixed diet, selecting in any season small morsels of highly nutritious foodstuffs; such a feeding style causes them to rely in the New Forest primarily, though not exclusively, on browse materials, particularly bramble, rose and other 'shrubby' species. Sika seem to fall somewhere between the two. Although their diet elsewhere suggests that they are preferential grazers, in the New Forest they take a considerable amount of browse and throughout the year the diet is about 30% grass, 40% heather, gorse and coniferous browse, although detailed proportions vary.

Such differences in feeding strategy between these three deer species are reflected in, and themselves reflect, actual anatomical differences in gut structure. In an extremely exciting investigation of the physiology of digestion of various ruminants, Hofmann and Stewart (1972) discovered a very clear relationship between feeding style and gut structure. Length of the gut, relative proportional size of different regions of the gut (in particular relative size of the four different chambers of the ruminant 'stomach'), even the fine structure of the lining of the rumen itself are all carefully adapted to the diet. Thus certain species have rather small rumina, and short hind guts, and the ruminal lining is resistant to high protein concentration: such animals were characteristically found to feed on small morsels of highly nutritious foodstuffs. By contrast, animals which feed unselectively, ingesting large quantities of relatively indigestible foodstuffs, require a very large rumen: reliant for the bulk of their nutrient supply on the breakdown of the structural carbohydrates of plant cell walls by symbiotic bacteria within the rumen, they must provide a huge fermentation chamber for this process. Absorptive linings to the gut walls are adapted to low nutrient levels and are rapidly damaged by 'concentrated' foodstuffs.

Recognition of these structural and anatomical differences, and of their precise adaptation to foraging strategy, led Hofmann and co-workers to draw up a comprehensive classification of large ungulates — based on their ruminal structure — as 'bulk-feeders', 'concentrate-selectors' and 'intermediate feeders' (Hofmann and Stewart 1972). On such a scheme, roe are expected to behave as true concentrate-selectors, fallow as bulk-feeders, and sika as just more 'intermediate' than the pure grazers (Hofmann *et al.* 1976; Hofmann 1982): predictions amply borne out by the ecological analyses of feeding style presented here.

Despite these overall differences in diet and feeding 'strategy' between the three deer species, however, many types of forage are taken by all three species, and there is in fact considerable dietary overlap at some/many times of year. The different deer diets *also* overlap with the foods taken by cattle and ponies in the same area (Chapter 4). How serious is this overlap — and is there evidence for competition between the five herbivore species?

We may examine the degree of overlap — and potential for competition — using the same analysis of 'niche overlap' developed in comparison of diet and habitat use between cattle and ponies on page 85. Here, we calculate overlap purely in terms of diet: for completeness, cattle and ponies are included in the analysis as well as roe, fallow and sika (Figure 6.3).

For most of the year little dietary overlap is observed between roe and the other two deer species. Overlap between roe and sika is consistently low, but in winter, when food is restricted both in quantity and variety, overlap between roe deer and fallow increases significantly. Jackson (1980) also noted that diets of roe and fallow deer within the Forest showed greatest overlap in winter and early spring. Concentrating on this period of the year on the assumption that, if there *is* any competition for food between the two species, it is likely to be at its most intense at this time, when food is shortest, he none the less concluded that widespread competition is unlikely to occur. Staple winter foods common to both species were coniferous browse, dwarf shrubs, fruits, bramble, rose and ivy, but there are clear distinctions in the relative importance that each food has in the total intake. By contrast, diets of fallow and sika deer show significant overlap throughout the year: both species are intermediate feeders on Hofmann's classification, and clearly both select the same types of food. Actual competition within the New Forest is unlikely, however: in those areas where sika deer are abundant, fallow are heavily culled specifically to reduce the potential for competition; the policy is successful and fallow numbers in these areas are extremely low.

Dietary comparison between the deer and the Common animals of the Forest reveals once again little overlap between the concentrate-selecting roe and the essentially bulk-feeding cattle and ponies. Overlap of cattle and ponies with the intermediate or bulk-feeding sika and fallow is higher, but is at its most intense from March to July — over the main part of the growing season when food is relatively more abundant. We should remember, too, that there is considerable habitat separation between the species. Although Common animals *do* gain access to Inclo-

		Cattle	Ponies	Fallow	Sika	Roe
	Cattle	*				
SPRING	Ponies	0.87	*			
(February-	Fallow	0.96	0.92	*		
April)	Sika	0.77	0.66	0.81	*	
	Roe	0.20	0.14	0.39	0.53	*
	Cattle	*				
SUMMER	Ponies	0.95	*			
(May-	Fallow	0.94	0.94	*		
July)	Sika	0.80	0.72	0.78	*	
	Roe	0.14	0.14	0.35	0.32	*
	Cattle	*				
AUTUMN	Ponies	0.96	*			
(August-	Fallow	0.86	0.88	*		
October)	Sika	0.90	0.80	0.79	*	
	Roe	0.20	0.17	0.45	0.31	*
	Cattle	*				
WINTER	Ponies	0.93	*			
(November-	Fallow	0.65	0.63	*		
January)	Sika	0.77	0.69	0.87	*	
	Roe	0.16	0.14	0.68	0.37	*
	Cattle	*				
ALL MONTHS	Ponies	0.96	*			
COMBINED	Fallow	0.92	0.91	*		
	Sika	0.84	0.75	0.84	*	
	Roe	0.18	0.15	0.43	0.37	*

$$\text{Overlap } O_{ij} = \frac{\Sigma p_{ia} \, p_{ja}}{[\, (\Sigma p_{ia}^2) \, (\Sigma p_{ja}^2) \,]^{\frac{1}{2}}} \qquad \text{(Pianka 1973)}$$

Figure 6.3: Niche overlap among New Forest herbivores for food use

sures, they occur there only in low numbers; really high densities are found only on the Open Forest. By contrast, the deer restrict much/ most of their activity to the Inclosures and make relatively less use of the commonable ground.

Further, these analyses of overlap should be interpreted with caution. High overlap does not necessarily imply competition; indeed, almost by definition, if high overlap *is* observed the animals must be exploiting superabundant resources. By converse, low overlap should not *necessarily* be seen as evidence of lack of competition; it could equally be that competition for shared resources has resulted in an ecological separation, that the competing species have been forced to adopt different diets to *avoid* competitive conflict. The fact that sika and fallow populations from different parts of the Forest show high dietary overlap reveals a high *potential* for competition. But, if fallow deer were allowed to establish in the sika areas, it is probable that analyses of the diet of that particular fallow population would reveal a great reduction in overlap.

In fact, the diets of horses, cattle, roe and fallow deer in the New Forest are much as would be expected elsewhere, suggesting that few 'adjustments' have had to be made to permit them to coexist within the Forest — and thus implying little direct competition. In sharp contrast, the diet of New Forest sika differs markedly from that which they appear to eat when allowed to 'do what they want' in isolation. Sika in Dorset and in five major forests in Scotland all have much the same diet (Figure 6.2). Only in the New Forest does the diet seem to change, with increased intake of browse and lower reliance on grasses. Nor is the direction of the change what one might expect in terms of the difference in habitat: the sika deer of Dorset and Scotland are animals of coniferous plantation and heathland; the New Forest offers a wider diversity of vegetation types with, in principle, *better* opportunities for grazing. Such an unexpected shift in diet may well then be the result of competition, and we suggested above (page 125) that in the New Forest there may indeed be real competition for forage between sika and the Forest ponies.

Chapter 7
The Pressure of Grazing and its Impact Upon the Vegetation

Earlier sections in this book have described various facets of the use of the available vegetation by the different Forest herbivores, in discussing patterns of habitat use, activity, feeding behaviour and diet. Each such usage exerts a certain pressure upon the vegetation. In this section I wish to try to pull together all the various uses of the Forest made by the different animals, in order to assess total impact on the vegetation.

We have already discussed the changing pressures of grazing on the Forest since its 'creation' in the eleventh century, and it is clear that fluctuations in numbers — and dominant species — of grazers over the years have had a marked effect on the Forest vegetation (woodland regeneration for example has been able to occur only sporadically — at times when grazing pressure was unusually low — a fact reflected now in a curious age structure of the mature trees of the Forest woodlands, page 153-4).

In the early history of the Forest, fallow deer were undoubtedly the dominant herbivores. Records are difficult to interpret, but through the seventeenth, eighteenth and nineteenth centuries numbers seem to have been steady at between 6,000 and 7,500 animals (page 88). Following the Deer Removal Act of 1851, populations were decimated: by 1900, censuses suggested a total count of only 200. Although the population has recovered from this time to its current maintained level of around 2,000 animals, the dominating grazing impact upon vegetation has clearly been that of commonable stock (Chapter 4). Densities of these Common stock have also fluctuated: Figure 7.1 summarises the changes over the past 30 years.

Estimates of numbers of grazing animals on the Forest as a whole — or even estimates of biomass — while they may highlight changes in grazing pressure over the years, are not a particularly sensitive reflection of the actual pressure sustained by the vegetation itself. The effect

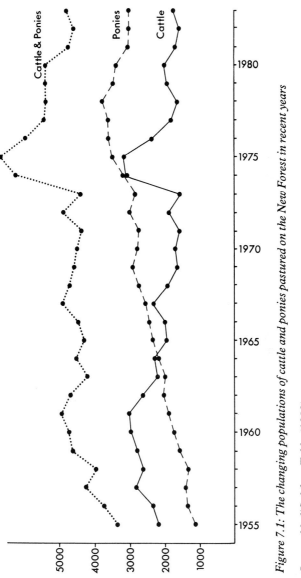

Figure 7.1: The changing populations of cattle and ponies pastured on the New Forest in recent years

Source: Modified from Tubbs (1982).

of absolute numbers is channelled through differential patterns of habitat use, differential use of different community types for forage, in determining actual grazing pressure upon the vegetation. It is clear from Chapters 3 and 5 that patterns of habitat use both by cattle and ponies and by the various deer species within the Forest are extremely complex, and that animal pressure is thus by no means distributed evenly over the environment. Certain vegetation types are but rarely used; others are particularly favoured for shelter, and thus subject to trampling; and others again are preferred feeding sites, subject to pressure by grazing and trampling. Thus, certain communities receive little herbivore pressure, whereas favoured sites may show extremely heavy use indeed. Yet the effects of these herbivores upon the vegetation are clearly going to be influenced not only by animal density on that particular vegetation type, but also seasonality of that pressure (a given grazing pressure at a particular time of year may have a very much greater or lesser impact than would that same pressure at some other stage of the vegetation's annual cycle). A more objective measure of animal pressure would take account of the temporal and spatial patterning of community use by the Forest herbivores to assess more precisely herbivore pressure on a particular area. Our studies of habitat usage by the different Forest animals enable us to arrive at such an assessment of grazing pressure, to translate a blanket figure of 'number of herbivores over the total Forest area' into a more meaningful figure of *intensity of use of particular vegetation types at particular times of year*, at least for the present day. We may calculate current herbivore pressure on an area quite precisely: in terms of animal-minutes, or animal-grazing-minutes, per unit area per unit time. These figures thus offer an objective measure of actual intensity of use of different vegetation types.

The importance of such a measure, in allowing us to quantify *effective* pressure of grazing animals, may clearly be seen if we remind ourselves of the pattern of pony use of different vegetation types within the Forest (Figure 3.2). This illustrates very clearly by way of example that intensities of use of different vegetation types are by no means equivalent. Heathlands, acid grasslands and bogs sustain the lowest pressure year round. Woodlands (particularly deciduous woodlands) are used consistently throughout the year for shelter, as well as feeding. But clearly the most heavily used communities are the various improved grasslands, which sustain at times an extremely high level of use. In fact, the ponies, as preferential grazers, spend nearly half their time on these grasslands over the whole year. The proportion of pony observations (as a percentage of total numbers observed on all communities) never drops below 35% in any month. The same picture, somewhat intensified, results from analysis of pony feeding observations. Thus, nearly 50% of all pony grazing pressure for the Forest as a whole is in fact concentrated on a mere 1,460 ha of grasslands. Such patchy use of different communities is also observed among the other herbivore species and emphasises the importance of analyses of habitat-use patterns in studies of grazing pressure. Figures for 'animal-minutes per unit area per 24

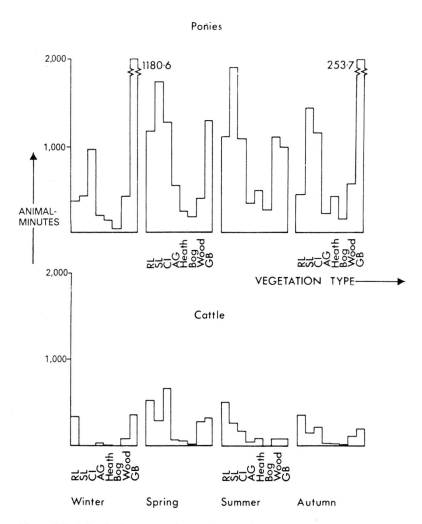

Figure 7.2: Animal-minutes sustained per hectare in an average 24-hr period

Table 7.1: Herbivore pressure: animal-minutes sustained per hectare in a 24-hour period (cattle and ponies only)

	Jan	Feb	March	April	May	June	July	Aug	Sep	Oct	Nov	Dec
Re-seeded lawn	201	888	763	1,978	2,007	1,575	1,793	1,421	717	960	634	838
Streamside lawn	362	745	2,598	1,898	2,646	1,611	2,590	1,883	1,195	1,798	1,488	141
Commoners' improved grassland	1,062	464	2,382	1,586	1,964	700	1,627	1,410	950	1,687	1,288	1,326
Acid grassland	190	412	48	675	1,074	274	480	329	338	98	356	60
Dry heath	13	56	34	58	78	130	60	64	66	48	116	40
Wet heath	81	81	177	88	164	119	197	139	130	76	73	30
Regenerating heath	33	—	680	53	266	406	283	189	224	395	314	159
Bog	18	74	124	151	345	242	258	313	276	72	151	36
Woodland glade	55	2,987	66	857	304	719	1,521	756	1,853	829	2,202	1,531
Deciduous woodland	329	594	476	350	581	230	580	264	385	425	237	149
Coniferous woodland	—	—	167	419	70	1,116	930	295	674	—	349	543
Gorse brake	11,455	13,091	1,933	2,259	702	1,870	624	883	936	2,961	4,026	11,963

hours' have been calculated for cattle and ponies on the major vegetation types of the Forest, and these highlight the tremendous pressure
which some of these communities may sustain (Table 7.1, and Figure
7.2).

An alternative approach to estimating pressure of animal usage, this
time restricted to consideration purely of actual *feeding* pressure, is in
analysis of above-ground vegetation production and percentage animal
offtake in each community. Table 7.2 presents data for this comparison
of vegetational productivity and offtake for a number of the more
favoured communities of the New Forest.

Measurements of forage standing crop, estimated above-ground
productivity and animal offtake were made over a period of two years in
most of the major vegetational communities of the Forest. In the first
year of analyses (1977-8) measurements of standing-crop biomass, production and offtake were made for entire communities (the complete
mixed sward of grasslands, or complete species mixture of bogs or
woodland floor); such measures, while relating closely to total vegetational productivity, combine forage and non-forage species as they
occur in the sward. Later analyses (1978-9) repeated these measurements, but extended them to examine in more detail the dynamics of
particular individual forage species.

The productivity and offtake of available plant material within *grassland, heather,* and *bog* communities were estimated using temporary
exclosures to prevent grazing (Plate 13). In this method the amount of
forage material present per unit area (standing crop) is assessed by cutting and weighing samples, both when the exclosures are erected and at
the end of the exclosure period. Thus:

$$\text{Production} = \begin{array}{c}\text{Final standing crop} \\ \text{(inside pen)}\end{array} - \begin{array}{c}\text{Initial standing crop} \\ \text{(outside pen)}\end{array} \quad (1)$$

$$\text{Offtake} = \begin{array}{c}\text{Final standing crop} \\ \text{(inside)}\end{array} - \begin{array}{c}\text{Final standing crop} \\ \text{(outside)}\end{array} \quad (2)$$

The length of time for which the exclosure is left is clearly critical.
Vegetation inside the pen is of necessity protected from grazing; yet the
vegetation it is supposed to represent normally grows under continuous
grazing pressure. With the relief from grazing afforded by the exclosure
(necessary for us to take any samples at all), we cannot be sure that the
pattern of growth of the ungrazed vegetation within accurately reflects
the growth pattern of the vegetation under its normal regime of heavy
grazing. Relief from grazing must therefore be for as short a period as

Table 7.2: Total annual production[a] and offtake in the different vegetation types (figures in g/m²)

Vegetation types	Growing-season production		Annual offtake		Offtake as % of production		
	1977/8	1978/9	1977/8	1978/9	1977/8	1978/9[b]	
Re-seeded lawns	349	226	221	296	63.3	131.0	Graze types
Streamside lawns	473	492	432	523	91.3	106.3	
Commoners' improved grasslands	321	329	313	371	97.5	112.8	
Acid grassland	190	158	132	250	69.5	158.2	
Bog	549	—	475	—	86.5	—	
Holly	—	27.3	—	1.9	—	6.8	Browse types

Notes

[a.] Production is measured over growing season only.[b.]

[b.] 1978/9 figures for offtake/productivity are frequently > 100%. Offtake during this period of very heavy grazing did indeed exceed production, and standing crop declined during the year.

Note, however, in addition that measurement of offtake as loss of standing crop attributes all such loss to animals. In practice, loss of standing crop may also include a component from vegetation senescence.

possible, yet cannot be too brief or the growth of vegetation within the pen is too small to measure. In practice, pens on grasslands, on bogs and on woodland floors were sampled on a two-monthly rotation; heathlands were clipped every six months.

Some care is needed in interpreting the results from these grazing exclosures. The problems of altered growth patterns as soon as grazing pressure has been removed have already been discussed, but, in addition, an underlying assumption of the method is that all changes in standing crop are due either to growth of the vegetation or to grazing. During the growing season this is probably a reasonable premise, but in the winter months the standing crop may decline owing to death and decomposition of plant material; a negative production value is quite possible and merely reflects this decline. Provided the loss in standing crop owing to death and decomposition is the same inside and outside the pen, then offtake can be calculated as in equation 2. In fact these losses are probably not the same, especially where there is trampling and disturbance outside the exclosure, and estimates of offtake may thus be inaccurate over the winter period.

Different methods need to be employed in estimating standing crop, production and offtake of shrubs and trees, since heterogeneity of the 'sward' at this size-scale means that no one site can be selected as an exact 'control' for any other area of woodland, however large the sample exclosure. Accordingly, the density per km of each of the common tree species within Forest woodlands was assessed by counting the number of trees of each species along a number of survey transects. Then, several individual trees were selected of each species, and the available browse (all material within 2 m of the ground) was clipped with secateurs to determine the weight of available browse per tree and the proportion of twigs showing browsing damage. Samples were taken in March and October: October measurements (at the end of the growing season) give a first estimate of available standing crop, and a measure of growing-season offtake; measurements in March offer confirmation of standing crop, and allow estimation of winter offtake. Actual offtake, by weight, was more difficult to determine. Each sample clipped was sorted into browsed and unbrowsed twigs; and it would appear simple to determine the mean weight of an unbrowsed twig and the mean weight of a browsed twig, and then calculate weight of offtake by difference. There is, however, such variance in the weight and length of the shoots that there is in practice little difference in mean weights of browsed and unbrowsed samples. Accordingly, unbrowsed twigs were sorted into size categories (on the basis of basal diameter) in an attempt to reduce the variance within each sample. *Then,* when variation around the mean weight of each class of twig was reduced to an acceptable level, weight of browsed samples could be compared with the weight of the unbrowsed shoots of an *equivalent size.*

A summary of results for forage production and offtake of the major communities and of the main forage species is presented in Tables 7.3 and 7.4. These data may be used in examination of animal feeding patterns; we have already seen that grazing by ponies and certain of the Forest deer is closely related to the timing of maximum production of each forage type. Results may also be used to investigate the effects within a given vegetation type of different intensities of grazing (as offtake) on productivity (page 163). In the present context, the data are used to assess grazing intensity through relationship of offtake to vegetation production. This relationship of offtake to forage productivity is widely applicable, in offering an independent estimate of grazing pressure in any system.

It is immediately apparent from Table 7.2 how high a level of offtake is sustained by the various Forest grasslands. Offtake in most graze communities is extremely high, amounting to nearly 100% of above-ground production for the year as a whole. Even during the summer growing season, month by month, 80-100% of the productivity may be removed at once by the grazing animals. Offtake of heather and woodland browse is far lower in proportion to their availability. (Data for holly are presented here since it is by far the most abundant browse species in woodland and has already been shown to be highly palatable to most of the Forest's herbivores: Chapters 4 and 6.)

Offtake figures in graze communities are tremendously high. To some extent such figures may be overinflated. As we have already noted, the method by which offtake is measured — effectively by change in standing crop — does not differentiate between losses due to grazing and those due to senescence and decomposition. During the growing season it is probably fair to assume that the majority of vegetation losses *are* due directly to grazing, but in the winter, when the vegetation is no longer growing rapidly and may be suffering considerable die-back, such an assumption may be unjustified. However, even if we recalculate the values of Table 7.2 for the main grassland types, but using data from the main growing season only (April-September), offtake during those months alone is still seen to be a *massive* proportion of production (Table 7.5). And we *know* that offtake calculated in this way is an underestimate: animals continue to graze in these same vegetation types throughout the year, and grasses and other

Table 7.3: Above-ground production of major vegetational communities and forage species. Figures, in g/m² are for the 1978-9 season

Month	Vegetation										
	RL	SL	CI	AG	Bog	WG	Ilex	Molinia[a]	Juncus[a]	Rubus[a]	Ulex[a]
January	—	13.5	9.2	6.5	13.9	—					
February	0.9	30.0	16.1	11.7	17.0	—					
March	5.7	25.3	0	0	51.2	—					
April	12.2	27.5	0	0	84.7	—					
May	27.1	49.6	46.0	8.6	95.5	—		May-Sept 226	May-Sept 1,228	May-Sept 318	May-Sept 722
June	39.0	72.0	78.9	26.2	118.0	28.4	24.4				
July	59.6	129.0	81.2	45.2	111.0	47.2					
August	50.3	117.0	65.5	33.9	42.4	26.7					
September	26.5	77.3	28.1	26.9	16.7	3.1					
October	5.2	43.7	6.3	22.7	11.6	—					
November	—	33.4	—	—	5.2	—					
December	—	38.3	—	16.8	—	—					

Note:[a] Results for these single-species samples are calculated per m² of plant area only. (Figures for whole communities are g /m² total ground area.)

Table 7.4: Monthly offtake of major vegetational communities and forage types 1978-9 (g/m^2)

	Vegetation							Molinia[a]	Juncus[a]	Rubus[a]	Ulex[a]
Month	RL	SL	Cl	AG	Bog	WG	Ilex				
January	15.3	46.2	14.0	14.3	—	9.7	Oct→	0	36.3	37.1	Oct→
February	11.1	40.9	12.5	7.8	—	6.5	April	0	28.6	16.1	April
March	8.7	33.8	6.7	3.7	—	2.3		0	9.7	8.4	
April	14.5	43.5	14.3	18.4	—	0.4	1.9	3.2	23.3	4.2	364.0
May	27.0	54.8	42.0	20.6	—	4.9		16.2	106.0	9.5	
June	38.7	64.5	61.5	13.7	—	13.2	April→	26.8	35.5	15.6	April→
July	54.4	94.5	64.2	22.1	—	8.8	Oct	35.1	187.0	19.7	Oct
August	48.1	71.5	50.1	37.9	—	11.9	0	20.7	132.0	20.8	0
September	29.3	42.7	36.0	54.0	—	16.8		4.8	116.0	20.5	
October	19.9	41.7	23.9	40.9	—	20.9		1.9	61.2	20.2	
November	20.2	24.6	13.7	16.3	—	10.9		0	29.6	27.5	
December	21.7	19.9	12.1	11.5	—	9.1		0	25.4	36.3	

Note:[a]. Results for these single-species samples are calculated per m^2 of plant area only.

Table 7.5: Total production and offtake of Forest grasslands during the growing season of 1979 (figures in g/m^2)

Vegetation types	Growing-season production	Growing-season offtake	Offtake as percentage of production
Re-seeded lawns	215	212	98.6
Streamside lawns	472	372	78.8
Commoners' improved grassland	310	268	86.5
Acid grassland	151	167	110.6

ground vegetation contribute a significant proportion of the diet of both cattle and ponies outside the immediate growing season (Chapter 4 and Figures 4.4, 4.5).

Such a calculation, however, makes another important point. We noted that the *timing* of defoliation was perhaps as important a factor to the plant as the actual amount (page 136). Yet it is clear from Table 7.5 that, in the grasslands at least, the majority of material produced is removed at the time, during the actual growing season. While offtake from heathlands and woodland browse is restricted to a relatively short period, and for the most part outside the growth period, close examination of data for grasslands reveals that offtake is more or less continuous. Not only is most of the growing season's production removed during the growing season, but the majority of it is removed *at once* — and on most communities figures for production and offtake balance even within *months* (Putman *et al.* 1981).

The Impact of Grazing upon the Vegetation

The effects of this tremendous grazing pressure are obvious. At a purely qualitative level one merely has to glance at the close-cropped turf of the Forest lawns, or to take in the absence of understorey in the Forest woodlands — and the clear browse line on the forest trees demarking the level to which the browsing ponies and deer can reach the foliage (Plate 15) — to appreciate just how significant the grazing pressure must be in the ecological balance of the Forest ecosystem. The influence on both structure and species composition is so obvious that it hardly needs quantification, but closer examination reveals even more wide-reaching effects.

The potential effects of heavy grazing upon vegetation discussed in the first chapter of this book may be summarised as:

1. Change in species composition through loss of graze-sensitive species, or encouragement of graze-resistant species, and through changes in nutrient availability.
2. Change in physical structure: grazing results in a loss of structural diversity.
3. Alteration to patterns of nutrient cycling both within communities and in transfer of nutrients from one community to another.
4. Stimulation or suppression of vegetation productivity.

Now that we have introduced the herbivores themselves, we shall examine the evidence for each of these various effects in the New Forest grazing system.

Changes in Physical Structure

Graze Communities. The patchy turf of Forest grasslands boasts no plant material higher than a few millimetres — apart from the occasional stem of ragwort! (Plate 14). Such grasslands clearly lack many of the possible structural layers of mature, ungrazed grasslands. On heathlands, too, the effects of heavy grazing are clear in their reduction of structural diversity. Heathlands in the south Hampshire area support dense stands of purple moor-grass (*Molinia caerulea*) among the *Calluna* and *Erica* heaths. Outside the Forest boundary the tall flower spikes of *Molinia* tower above the canopy of the heather plants and provide a whole additional structural element within the vegetation. Inside the Forest this whole stratum is missing. *Molinia* plants are just as abundant within the heathlands, but *Molinia* is an important component of pony diet (Chapter 4) and the plants are always heavily grazed — kept well below the heather canopy. This loss of a structural element within the heathland community is dramatically illustrated on the boundaries of the Forest where a simple fenceline divides a single heathland. On one side — inside the Forest — ponies graze undisturbed; from the area outside the Forest boundary, ponies and cattle are totally excluded. The geology, soil type and weather conditions are identical on either side of the fence; the only difference is the intensity of grazing — and the vegetation itself.

But, while grazing appears to reduce the structural diversity of vegetation on the vertical scale, it actually increases diversity of the sward itself, promoting discontinuity and patchiness within a homogeneous community. Perhaps the clearest example of this last effect is the recognition within the improved grasslands of the Forest of distinct sub-communities corresponding to those areas where animals graze and those where they defaecate (see page 58-9). In their use of such grasslands, the Forest ponies establish traditional latrine areas to which they move to defaecate. No large herbivore feeds in the immediate vicinity of fresh dung, and as a result there is a mosaic established within these grasslands of short-grass areas where the ponies graze, and relatively ungrazed, but nutrient-rich, latrine patches.

This effect might not be expected to persist in a multi-species system; but the action of the other herbivores does not counteract this mosaic pattern. The other major herbivores of these grasslands are the cattle; their mouthparts are such that they cannot feed on the close-cropped turf of pony-feeding areas, but are perforce restricted to latrine areas. (Since they have no traditional latrines of their own, but defaecate where they happen to be, the dung remains in these pony-latrine areas and is not returned to the short-turf patches.) As a result, owing to the grazing and eliminatory behaviour of the herbivores, distinct sub-communities are established and maintained within the improved grasslands which differ both in species composition *and* in overall intensity of grazing; these two factors contribute to a distinct difference in physical structure (although a subtle one: areas of pony grazing have a maximum turf length of about 5 mm; vegetation in latrine areas rarely exceeds 2 or 3 cm). Although these sub-communities are most apparent on the Forest lawns, we have data to suggest that the same latrine behaviour may lead to heterogeneity in other communities as well — certainly on acid grasslands and heathland, where the development of gorse brakes seems to correlate well with patterns of animal elimination.

These effects on vegetational structure might strictly be considered effects of eliminatory behaviour rather than direct effects of grazing. But grazing patterns, and different intensities of use of different areas within a community, can also lead to the development of structural diversity. In Chapters 3 and 5 we have shown that grazing pressure by deer and domestic stock is not evenly distributed across all vegetation types, but that some are more favoured and subject to greater intensity of use than others. This holds even *within* a single vegetation type: ponies grazing pasture on abandoned farmland in the Netherlands show clear preference for different parts of their range, despite apparent vegetational uniformity (Oosterveld 1981). Oosterveld has shown clear differences in the scale of use of a single 100-ha grassland system by free-ranging Iceland ponies (at a density of 0.3 per hectare) to the extent that 3% of the area is heavily used, 70% only moderately or highly used, and 27% is subjected to a very low level of impact or never used at all. Presumably, some areas are more sheltered than others, or are favoured for some other, perhaps social, reason; whatever the reason, distribution of animals is uneven — and the uneven grazing pressure that results leads ultimately to structural diversity within the habitat (Table 7.6).

Woodlands. In woodland communities of the New Forest, once again, some of the effects of animal usage on physical structure of the vegetation are immediately apparent. New Forest woodlands virtually lack *any* ground flora or shrub layer. The woodland floor is essentially bare, and indeed the whole structural 'layer' between ground level and 1.8 m — the extent of a pony's reach — is missing: most of the Forest woodlands display a marked browse line at this level (Plate 15). Under continuous browsing pressure, palatable shrubby species such as hawthorn, blackthorn and hazel are eliminated and fail to regenerate. Even species relatively resistant to grazing, such as holly or gorse, are heavily used: taller holly trees are thoroughly browsed up to the 1.8-m browse line and have little vegetation below this level; shrubs of both holly and gorse which fall entirely within the reach of the herbivores are severely stunted and 'hedged' by the continuous browsing (Plate 16). At the ground level, brambles, ivy and other low vegetational species are completely eliminated; the only species which gives any structure at this

Table 7.6: The effects of uneven grazing pressure by ponies in creating different structural patches within an area of disused farmland

| | Development of structural diversity within the sward (%) | | | |
Grazing intensity	Short-grazed (25 cm)	Rough stands (25 cm)	Bushes, woody plants (2.5 m)	Wood (2.5 m)
Heavy	100			
Moderate	80	20		
Light	10	50	30	10
Very light	5	10	25	50

Example: A moderate use of the area means that you get 80% short-grazed and 20% rough vegetation as a result; at moderate stocking rates, 60% of the area is moderately used.
Data from Oosterveld (1981).

level is bracken (*Pteridium*), which, although eaten by the ponies at certain times of year, is not particularly palatable.[1]

A direct analysis of the overall effects of grazing on woodland structure may be found in a unique experiment carried out within one small area of Forest woodlands. In 1963, 11.2 ha of mixed (oak/beech) woodland in Denny Lodge Inclosure were fenced to create two separate, but adjoining, compounds of approximately equal area. All grazing animals were removed from one pen; in the other, a population of fallow deer was maintained at known density (averaging one per hectare over the years). The pens were surveyed after six years and again after 15 years, to consider changes in vegetational character in the two areas (Mann 1978). Strictly speaking (since both pens were originally heavily grazed), differences between the two pens should be considered due to secondary succession within the ungrazed pen on relief from grazing, rather than viewed the other way round. Since the vegetation of the ungrazed pen is secondarily developed from an area once heavily grazed, it cannot be assumed equivalent to the primary vegetation cover of the area before grazing; it is therefore not legitimate to use a comparison of the two pens to discuss in detail the effects of grazing in the grazed pen — because the vegetational characteristics of the ungrazed plot are not a good control. None the less, results do clearly demonstrate the effects of *continuation* of grazing in preventing secondary succession, and thus offer some indication of the effect of *current* grazing on community structure in preventing this successional development elsewhere in the Forest.

The two 5.5-ha pens were compared six and fifteen years after the exclusion of all the large herbivores from one pen. In the first survey (intended as an interim measure to evaluate some of the more major differences between the areas and to offer an earlier 'time point' to assist with interpretation of a later, more complete, analysis) records were taken of tree numbers and species composition of each pen and biomass of ground vegetation. The second survey repeated these basic measurements, and in addition recorded species diversity and structural diversity of the vegetation in each pen. Although carried out on only a relatively small scale, the study was tremendously detailed and its results are worth considering here in some depth.

A 50-m grid was established within the area by recognised survey

---

[1] Such an impression, however, striking, is none the less subjective. A more formal analysis of the structure of the 'ground layer' in woodlands inside and outside the Forest boundary, based on hemispherical photographs, is presented by Hill (1985).

techniques. Within each of the 2,500 m² squares resulting, a 1-m² quadrat was located. Twenty such quadrats were identified in this way in each pen and these were used as sampling points. Vegetational parameters recorded in each quadrat were:

a) Tree numbers   (6 and 15 years)
b) Species composition (6 and 15 years)
c) Biomass   (6 and 15 years)
d) Structural diversity (15 years only)
e) Species diversity   (15 years only)

In the 1969 survey all the trees in each pen were plotted on a map (scale 50″ to the mile) by reference to the grid, and a list was compiled of the species and the numbers of each species occurring in each pen. In 1978, an analagous result was obtained by establishing a circular quadrat (radius 10 m) centred on each quadrat, and recording all the trees within that area. Species composition of the ground vegetation was assessed by counting all the plant species present in each of the 40 quadrats. In 1978, relative abundance of the various species within the quadrats was recorded in addition, using a 'pinframe'. (The three-dimensional pinframe is an extension of the point-quadrat method of sampling whereby the recording of the number of hits on a species by pins lowered into the vegetation allows a number of quantitative measures to be extracted from the records.)

In 1969, and again in 1978, the standing-crop biomass of each quadrat was clipped as close to the ground as possible and the vegetation was weighed fresh, as a crude estimate of structural diversity based on vegetational density. Vegetational structure may be more precisely defined, however, in terms of relative density of vegetation and the way this changes with height above the ground, and in 1978 measures of actual vegetation densities at various heights were made, once again employing a pinframe. The pinframe used for this experiment consisted of a single bar, 1 m long, which could be moved across the quadrat in stages (in this case 10 cm). Holes were drilled in the frame at 10-cm intervals and the pins were lowered in 10-cm stages. Pins terminated in a needle in order to approach as closely as possible the requirement of a 'true point', and records were made of the contacts between the pin tip and the herb species for each point on the matrix. In this instance detailed records were made of the number of hits per species per quadrat, at every 10-cm stage above the ground, giving an index of vegetation density at 10-cm intervals up to 1.5 m.

Published regressions relating the girth at breast height (or sometimes at ground level) to timber volume, leaf weight and area, and leaf
and timber production (Whittaker and Woodwell 1968) were used to
estimate vegetation densities above 1.5 m. Finally, measures of species
diversity of the ground flora of two pens were made during the 1978 survey. (These are discussed later, page 155-6).

The gross morphological and vegetational differences between the
areas are very clear. In one pen there is a distinct browse line and vision
is relatively unimpeded, easy progress being possible in any direction.
The regeneration in the ungrazed pen, on the other hand, has now
reached such a stage that passage through it is difficult; the area can
accurately be described as a thicket, and it is full of young, very vigorous
saplings. There is no evidence of a browse line, and ground vegetation
reaches up into the canopy of the trees, forming a dense and deep layer
of plant material.

Fifteen years after fencing, there were shown to be 35 times as many
trees in the ungrazed pen as in the pen still grazed by fallow deer,
although the difference between the compounds is not significant until
the regenerating trees are included in the analysis. This last point shows
the equivalence of the two areas before enclosure, and stresses the effect
of grazing in prevention of such regeneration: there are in fact *no* tree
seedlings or saplings recorded in any of the samples taken from the
grazed pen (Table 7.7). (Regenerating trees in the ungrazed pen — that
is those included here being less than 1.5 m in height — were in fact more
common in the 1969 results, e.g. *Pinus sylvestris, Pseudotsuga menziesii,
Larix* sp., but figure less in the later study as they have grown out of that
category. The 1978 data do include an imbalance of *Salix* saplings, with
again more in the ungrazed pen.) Canopy *structure* also differed markedly between the two pens. The canopy in the ungrazed pen was not
distinct and reached down into the bracken and bramble clumps; there

**Table 7.7: Total number of trees per hectare in the Denny pens
exclosures**

	Pen	Trees/ha	Mature trees	Regenerating trees
1969	Grazed	125		
	Ungrazed	106		
1978	Grazed	188	188	0
	Ungrazed	6,893	250	6,643

was no indication of a browse line. The grazed pen had no canopy vegetation below 1.5 m and the browse line was very distinct.

Figures for standing-crop biomass of *ground vegetation* are also higher in the ungrazed than in the grazed pen, as might be expected. Results from both 1969 and 1978 agree on this point. The more detailed analysis of 1978, however, suggests that this difference is barely significant (Table 7.8). Vegetational density can also be considered in terms of volume, or 'bulk'. In comparison of grazed and ungrazed pens in terms of the bulk of ground vegetation, pinframe results were used to calculate the total number of pinhits from each set of 20 quadrats, and then to find the mean number of hits per quadrat. The results are given in Table 7.8, where a comparison of means indicates that the areas are not significantly different in this parameter.

Structure is, however, three-dimensional; differences in physical structure of the vegetation between the grazed and ungrazed pens should properly consider 3-D *distribution* of this vegetational bulk. Mann presents an index of structural diversity using pinframe data. This index considers the proportion of plant material occurring at a particular height above the ground in any quadrat, and is derived as the number of pinhits at a particular height (chosen to suit the circumstances) as a percentage of the total number of hits for the quadrat, as:

$$\frac{\text{Number of pinhits at height x}}{\text{Total hits for the quadrat}}$$

In this study, the index was calculated at 0, 10, 50, and 100 cm above ground in each quadrat. The results reveal a significant disparity between the pens for the lower levels, 0 and 10 cm, but, interestingly, *not* at the higher levels, 50 and 100 cm (Table 7.9).

Table 7.8: Mean standing crop of ground vegetation in Denny pens experiments (figures in g/m²)

	1969	1978
Grazed pen	138	495
Ungrazed pen	605	687

Vegetation density from bulked pinhit data (1978 data)

	Grazed pen	Ungrazed pen
Total pinhits per pen	1221	857
Mean hits per quadrat	61.05	42.85

Table 7.9: Mean vegetation-density index Denny pens experiments (1978: see text, page 152)

Height (cm)	Grazed pen	Ungrazed pen
0	16.5	2.5
10	21.9	11.6
50	7.3	10.4
100	1.7	3.0

Results from this study highlight the differences in vegetational composition and structure between woodland areas free of grazing animals and those maintaining a high density of large herbivores. Differences in structure — and actual vegetational bulk in terms of pure biomass or bulk of material — are apparent both in the ground flora and in the woody vegetation. Much of the difference between the grazed and ungrazed area results from the massive regeneration of tree species which has occurred in the area free of grazing; results emphasise the lack of such regeneration in grazed areas. This suppression of regeneration may have a direct effect on physical structure of the vegetation, but in the long term has an even more significant effect — on the population age structure of the Forest trees.

Were regeneration to be prevented for ever, the woodlands would of course ultimately disappear altogether as the existing trees aged and fell and no new trees could establish in replacement. Intensity of grazing pressure fluctuates over the years, however, and at times of relatively lower herbivore numbers some regeneration may occur — enough to guarantee continuation of woodland cover. But such regeneration is spasmodic, and can occur only over brief periods when herbivore density is low enough to permit the survival of at least some of the seedlings. This alternation of periods of heavy and lighter grazing pressure, with resultant alternation of periods when seedlings may and may not establish, has a curious effect on the age structure of the populations of trees within the Forest woodlands.

Deciduous woodlands in the Open Forest in fact consist of trees of only three major age classes: those which germinated during, respectively, 1650-1750, 1860-1910 and 1930-1945 (Peterken and Tubbs 1965). These periods correspond to times when numbers of grazing animals on the Forest were reduced. In the intervening periods, animal numbers rose so high that no further regeneration could occur.

The trees of 1650-1750 — Peterken and Tubbs' A-generation — are the last survivors of that period of history when the first Inclosures were made within the Forest for timber production (page 19). The evidence suggests that this A-generation of trees arose as the result of deliberate management when, by enclosure, establishment of trees by natural regeneration was permitted on sites at that time largely cleared of their former tree cover. (It is of considerable interest to discover that the 1650-1750 regeneration in the New Forest had its counterparts in other Royal Forests: Tubbs 1968.) Relaxation in the administration and management of the Forest, and a deterioration in control exercised over the Forest deer, spelled the end of this phase of regeneration in the middle of the eighteenth century.

Herbivore numbers remained high enough to prevent further seedling establishment until over a century later. The regeneration of 1860-1910 (and the establishment within the Open Forest of Peterken and Tubbs' B-generation) followed the killing of large numbers of Forest deer when the Crown relinquished its hunting rights on the Forest in 1851 (Chapter 2); that of 1930-1945 coincided with lowered populations of domestic stock over this period as a result of low market prices in the recession.

This picture is somewhat oversimplified, for, although regeneration ceased around 1920 at the end of the B-generation, herbivore grazing pressure was no higher at this time than in the 1880s — in the middle of the regenerative phase. What, then, finally finished the period of regeneration? Tubbs (1968) points out that an investigation of the rate of growth of B-generation hollies, by means of growth-ring counts at successive heights, showed that the period of rapid height growth ceased between 1910 and 1920. Seedlings established after 1910 were heavily suppressed, so it may be inferred that the canopy closed during those ten years. On sites where the B-generation holly had developed as a scatter of trees with an incomplete canopy, regeneration was found to have continued throughout the 1920s and 1930s. This and other evidence suggested that the B-generation largely ceased because the canopy itself prevented further regeneration (Tubbs 1968), rather than as a direct result of increased grazing pressure. But, if the closing of the B-generation canopy itself inhibited further regeneration, it remains necessary to explain why the outer margins of the woods did not continue to expand and why new areas of scrub and emergent woodland did not come into being on open, unshaded heathland sites at that time. In recent years there *has* been some encroachment of birch and Scots pine on areas of open heathland, despite the extremely high numbers of cattle and ponies on the Forest — regeneration, despite a grazing pressure which is probably higher than that experienced in the 1920s. The evidence here points to the development of heath-burning practices on an unprecedented scale in the past hundred years, which severely restricted the spread of woodland; reduction of this large-scale burning in recent years has perhaps permitted the more recent scrub encroachment observed.

Changes in Species Composition

Many of the changes in physical structure which result from the heavy pressure of herbivores on the Forest vegetation are accompanied by, or are indeed *due* to, actual changes in species composition within the plant community. Thus, the lack of the understorey shrub layer of woodlands is due to the selective eradication of species such as hawthorn, blackthorn, hazel and willow.

Even moderate levels of grazing may result in loss from a community of particularly graze-sensitive species. Under heavy grazing pressure, even moderately *resistant* species may be eradicated if they are particularly palatable and thus exposed to particularly high levels of use. As a result, any species which are particularly palatable or graze-intolerant are lost from the vegetational community. By the same token, species which are resistant to grazing (possessing physical or chemical defences) or graze-tolerant (by growth form and physiology able to withstand some degree of defoliation) may be encouraged. Grazing restricts the distribution and performance of potential competitors, allowing existing graze-tolerant or resistant species to expand, or new species to enter the community. Moderate or heavy levels of grazing can thus dramatically alter relative abundances of graze-sensitive and graze-resistant or -tolerant species within the community, altering its entire species composition.

Chris Mann's study of the Denny pens (pages 149-53) showed clearly the selective loss under heavy grazing of species such as hawthorn, blackthorn and hazel, on which we have already remarked. In the secondary succession permitted within a small enclosure after relief from grazing, all these species rapidly re-established themselves. Abundance after 15 years relief from grazing is shown in Table 7.10 by comparison with abundances in the adjacent, grazed, area. While less tolerant species such as these are quickly lost when exposed to heavy use, holly (*Ilex aquifolium*), a resistant species, shows considerable expansion under these conditions and has a far greater abundance and wider distribution within the Forest than would be expected by comparison with equivalent woodland areas elsewhere.

Similar changes in species composition can be observed in the ground flora. Plant species were recorded in each quadrat at Mann's

Table 7.10: Species composition of trees in two woodland enclosures, one grazed, one ungrazed, as mean number of trees per 10-m radius circle. Data from Mann (1978)

Species	Grazed pen	Ungrazed pen
Fagus sylvatica	2.54	22.4
Quercus sp.	1.75	12.9
Pinus sylvestris	1.0	43.25
Larix sp.	.35	6.3
Betula sp.	.25	65.25
Pseudotsuga menziesii	.15	13.6
Ulex europaeus		19.66
Ilex aquifolium		17.05
Crataegus monogyna		.96
Prunus spinosa		.6
Salix sp.		23.8

study-site on a presence or absence basis, and these results provide a list of 'species occurrences' for each compound. An index was devised to emphasise the differences beween the compounds and to 'weight' those species more prolific in one pen than the other. The index was calculated from

> % quadrats occupied in
> ungrazed pen − % quadrats occupied in grazed pen

and it has a range of values from +100 to −100. Those species with an index of zero occur equally in both regions; those with a negative index are more abundant in the grazed pen. More species of grasses were found in the grazed pen, a shift in community structure frequently noted in response to heavy grazing. Grasses grow from the base rather than the tip, and are thus more tolerant of grazing than many other species (page 181); small ground-covering herbs, however, were similarly distributed between the two pens (Table 7.11).

In almost all vegetation types within the New Forest it is possible to demonstrate similar effects. Re-seeded lawns, established in the 1950s with pasture seed mixes containing several species of herbage plants of which perennial rye-grass and white clover were usually the most important, show now a tremendous change in species composition. Current studies demonstrate almost total loss of many of the original, more palatable forage species, with increased domination of the areas by *Agrostis tenuis*.

Table 7.11: Index of species occurrences in grazed and ungrazed woodland pens (Denny pens experiment, see text, pages 155-6)

1969 survey	Index	1978 survey
	65	Rubus
Calluna	30	Calluna
Salix	25	Betula
Pinus sylvestris, Betula spp.	20	
Rubus	15	Hypericum, Rosa, Salix
Epilobium, Ulex, Pseudotsuga mensiesii, Ranunculus, Viola	10	Hedera, Ilex, Potentilla, Ulex
Crataegus, Larix sp., Erica, Oxalis	5	Agrostis stolonifera Anthoxanthum, Crataegus, Epilobium, Pinus, Teucrium
Molinia	0	Quercus
Euphorbia amygdaloides, Juncus conglomeratus, Pteridium, Ilex	−5	Fagus, Lonicera, Oxalis, Rumex
Deschampsia cespitosa, Carex sp.	−10	Pteridium aquilinum
Hedera helix	−15	Digitalis, Luzula campestris, Viola
Lonicera	−20	Molinia
Rosa, Luzula campestris	−35	Deschampsia cespitosa, Juncus spp.
	−45	Agrostis tenuis

About 20 areas of acid grassland, gorse and heathland (totalling some 350 ha) were ploughed and fenced during the Second World War, and cropped for cereals or potatoes. After the war these areas were re-seeded with grass and opened to stock, providing improved grazing for the increasing populations of commonable animals. The seed mixtures and fertilisers used varied considerably in different areas, with perennial rye-grass (*Lolium perenne*) cock's-foot (*Dactylis glomerata*) and Timothy (*Phleum pratense*) particularly common as grass mixtures, accompanied by heavy applications of lime, chalk and superphosphate. Red fescue (*Festuca rubra*) and crested dog's-tail (*Cynosurus cristatus*) were other commonly used grasses, and many swards contained white clover (*Trifolium repens*). In most cases establishment of the sward was successful; since the re-seeded lawns were 'thrown open' to stock, however (at various times between 1948 and 1953), their species compositions have changed considerably, with a gradual invasion of native species and almost complete elimination of all those that were sown. The present species composition of a number of such lawns is shown in

Table 7.12. In general, the species composition is now remarkably uniform, both within and between lawns, with a very characteristic and constant suite of species on all well-drained sites. As noted, *Agrostis tenuis* has established itself as the most abundant species, with a percentage cover ranging between 31% and 57% (mean 45%). Other species which were invariably present (mean percentage cover in brackets) were *Bellis perennis* (5%), *Hypochoeris radicata* (5%), *Plantago lanceolata* (4%), *Sagina procumbens* (3%), *Taraxacum officinale* (2%), *Luzula campestris* (3%), and *Trifolium repens* (3%). Thus the lawns are largely composed of low-growing species and rosette plants, well adapted to heavy grazing pressure and trampling. In addition a number of annual species are usually present, especially in the short-grass areas where they take advantage of small gaps in the sward. Among the most constant of these are *Trifolium micranthum*, *T. dubium*, *Vulpia bromoides* and *Aira praecox*.

We can actually examine the *timing* of these changes in species composition since the sites were first re-seeded. Pickering (1968) provides a detailed survey of the species composition of several lawns in 1964,

Table 7.12: Species composition of New Forest grasslands. Mean values of percentage cover are based on several samples; from six sites

	Re-seeded Lawns
Agrostis tenuis	45
Cynosurus cristatus	1
Festuca rubra	8
Lolium perenne	2
Poa compressa	2
Sieglingia decumbens	1
Bellis perennis	5
Hieracium pilosella	1
Hypochoeris radicata	6
Leontodon autumnalis	3
Lotus corniculatus	2
Luzula campestris	3
Plantago coronopus	1
Plantago lanceolata	5
Prunella vulgaris	1
Sagina procumbens	2
Taraxacum officinale	2
Trifolium repens	3
Other species	3
Bare ground	4

between four and 15 years after the various different sites had been thrown open to grazing. He showed how the number of colonising species increased with time and was highest in the oldest swards. For the first few years the sown species, particularly *Lolium perenne* and *Trifolium repens*, were fairly successful. In all cases the most abundant of the invading species was *Agrostis tenuis*, which showed a very rapid increase in abundance between ten and 15 years after seeding to achieve cover values ranging between 30% and 50%. The 15 years between Pickering's survey and our own saw the almost complete elimination of the sown species, and in most cases the present species composition has almost nothing in common with the original seed mixture (Table 7.13).

It is, however, not purely through actual grazing that large herbivores may change the species composition of their vegetational environment. Trampling has a direct impact; and both urine and dung, in causing locally marked changes in soil conditions, may also cause changes in floral composition. We have already discussed the mosaic of distinct sub-communities created on the Forest lawns in response to the eliminatory behaviour of the ponies. Patterns of defaecation by the ponies, coupled with a reluctance to graze near their own faeces, result in the establishment of distinct feeding and latrine areas on the Forest grasslands. Because of the shortness of the turf in areas grazed by ponies, the other major herbivores, cattle, are forced to graze in pony latrines. Their dung also accumulates within these latrine areas (page 58-9). Within the grasslands we thus find areas characterised by high nutrient input and reduced grazing pressure (cattle cannot crop the herbage as closely as ponies) and other areas under high grazing pressure and suffering continual loss of nutrients. Characteristic of these distinct conditions, each sub-community has its own, markedly different species composition. The most obvious differences are the presence of ragwort (*Senecio jacobaea*) and thistles (*Cirsium arvense, C. vulgare*) only on latrine areas; other differences are noted in Table 7.14.

Similar effects may be observed in different communities: heather plants (*Calluna*) are easily damaged by trampling and the shoots killed by contact with nitrogen-rich urine. Bristle bent grass (*Agrostis setacea*) and heaths (*Erica* spp.) are also species easily killed by urine. In contrast, species such as carnation sedge (*Carex panicea*), purple moor-grass (*Molinia caerulea*) and common bent (*Agrostis tenuis*), although they may be scorched by direct contact with urine, recover quickly and grow more strongly afterwards in the enriched soil. As a result, trampling and eliminatory patterns may also alter the species composition of, in this case, heathland.

Table 7.13: Comparison of species composition (percentage cover) of Longslade re-seeded lawn in 1958, 1963 and 1979. (Data for 1958 and 1963 from Pickering 1968; data for 1979, Putman *et al.* 1981)

	1958[a]	1963	1979
Agrostis canina	—	1.5	
Agrostis tenuis		47.8	51.0
Cynosurus cristatus		1.2	0.9
Dactylis glomerata	58	6.8	2.4
Festuca rubra			2.9
Holcus lanatus		1.1	1.2
Lolium perenne		2.0	2.5
Poa annua		0.2	0.7
Poa sp[b]		0.9	0.9
Sieglingia decumbens		0.6	0.6
Vulpia bromoides		5.3	0.1
Bellis perennis		4.7	6.3
Cerastium holosteoides		0.6	0.2
Hieracium pilosella		0.2	0.3
Hypochoeris radicata		3.2	3.8
Leontodon autumnalis		3.7	1.8
Luzula campestris		0.8	2.0
Plantago lanceolata		0.3	7.4
Plantago major		0.2	
Potentilla erecta		0.3	0.1
Prunella vulgaris		0.9	0.6
Ranunculus sp.[c]		0.5	0.4
Sagina procumbens		1.2	0.4
Senecio jacobaea		0.3	0.1
Taraxacum officinale		0.5	1.6
Trifolium dubium		0.4	0.3
Trifolium pratense	10	0.6	
Trifolium repens	32	12.8	3.2

Notes
a. Estimated from seed mixture.
b. Identified as *P. pratensis* in 1963, as *P. compressa* in 1979.
c. Identified as *R. repens* in 1963, as *R. acris* in 1979.

If such changes in species composition of vegetational communities caused by grazing, trampling, or patterns of elimination continue over a protracted period, we may observe gross shifts in community structure, even conversion of one entire community type to another. Continued use of woodlands with no respite for regeneration would result, in extremes, in the establishment of grasslands — as parkland savannas in

Table 7.14: A summary of the main differences in plant-species composition between 'short-' and 'long-grass' (latrine) areas of re-seeded lawns

Status	Species	Proportion of sites	% Cover of Latrine areas	% Cover of Short-grass areas
Confined to long grass	*Cirsium arvense* *Cirsium vulgare* *Senecio jacobaea*			
More abundant in latrine areas	*Hypochoeris radicata*	6/6	7.7	4.8
	Lolium perenne	5/6	2.2	0.3
	Trifolium repens	5/6	5.3	1.3
More abundant in short grass	*Poa compressa*	4/6	1.8	2.8
	Sagina procumbens	4/5	1.2	2.2

Africa are prevented from completing their succession to woodland by fire and the periodic ravages of elephant. On a shorter time-scale, however, we may appreciate more subtle changes in community structure. Urine scorch of heathland, discussed above, causes loss of certain base-intolerant species such as *Calluna*, *Erica* spp. and *Agrostis setacea*, while encouraging the growth of other species, e.g. *Agrostis tenuis*. Use of heathland areas bordering the Forest grasslands as latrine sites by the ponies thus encourages the outward spread of the grassland itself. Ponies feeding on the edges of grasslands regularly move off the areas into adjacent ones to defaecate. If acid-loving plants are eliminated in these communities and replaced by grasses, these areas will eventually be used for grazing; the continued grazing maintains the grassland sward and prevents return to a heathland vegetation, and the heathland is gradually converted to grassland. Similar factors influence the development of patches of gorse (*Ulex europaeus*) around the edges of the same grasslands (Putman *et al.* 1981).

The effects of large herbivores upon their vegetation, through grazing, trampling or elimination, can clearly dramatically influence the species composition of that vegetation. This has an immediate and dramatic effect on the functioning and development of these communities, for it alters not only the species structure of the community, but also its physical structure (section one here). We can begin to appreciate what a dominating role the large herbivores must play in the creation and maintenance of the many curious vegetational formations characteristic of — and peculiar to — the New Forest.

Alteration to Patterns of Nutrient Cycling and Productivity

Since the Forest supports such an enormous biomass of grazing herbivores, a large proportion of its available nutrient resources must, at any one time, be bound up in animal tissue. Nutrient cycles become tighter and the system becomes more fragile (Putman and Wratten 1984). In addition, soils and vegetation are nutrient-impoverished.

In addition to these overall effects, actual patterns of feeding and elimination by large herbivores alter the detailed pattern of nutrient cycling, and can result in gross shifts of nutrients around communities or from one community to another. A clear illustration of this may be drawn from the curious pattern of use of improved grasslands shown by ponies and cattle, mentioned earlier.

Establishment by the ponies of traditional grazing areas and different, latrine, patches within these grasslands leads to a gross shift of nutrients from grazed to ungrazed portions of the lawn. Nutrients are continually deposited in the latrines, continually removed from the grazed areas. Nor does the action of the other main herbivores (cattle) redress this balance, since they both feed and defaecate primarily within the pony latrines. As a result, there is continued transfer over the years of nutrients from areas grazed by ponies to latrine patches, and a continued impoverishment of the grazing. The most consistent differences are in potassium and phosphorus content of the soil, this being higher in latrine areas by a factor of some 1.2 (phosphorus) to 1.7 (potassium) (Putman *et al.* 1981). Organic-matter content of the soil in latrine patches is also consistently a little higher. Only where periodic flooding spreads the dung more evenly over the area (as on stream banks) is this continued gross shift of nutrients reversed. (This difference in nutrient status of the soil is a major contributing factor to differences in species composition between these different sub-communities, noted earlier.) Clearly, disruption of nutrient cycles, or continuous depletion of available nutrients in an area, will result in not only a change of plant-species composition, but a reduction of productivity. Grazing and eliminatory patterns may thus markedly affect patterns of vegetational production.

In addition, grazing itself may have more *direct* effects on vegetational productivity. Various studies in the literature (page 3) have established that grazing and browsing affect productivity of the vegetation exploited: with low levels of grazing stimulating regeneration and enhancing productivity, and higher levels of foraging suppressing growth by reducing effective photosynthetic area below an efficient threshold. Clearly, the extremely high levels of grazing in the New Forest, particularly on grasslands, must be having an affect on forage productivity directly, and not just through soil-nutrient impoverishment. Offtake in browse communities is only a relatively low proportion of production, and too low for us to detect any changes in productivity as a result. As we have seen, however, the various Forest grasslands sustain an extraordinarily high grazing pressure.

The effects of this grazing pressure have been examined in the natural situation, by comparing productivities of lawns under different grazing pressures. In addition, a number of experimental plots were set up, grazed *artificially* at different intensities. (Since different intensities of natural grazing will actually be reflected in the time interval between successive bites suffered by any one plant, clipping plots by hand at different time intervals simulates quite accurately different levels of use.) Plots were established in 'long-grass' areas (pony latrine) and on the shorter turf of pony-grazing areas, and clipped fortnightly, monthly and at two-monthly intervals. Plots in latrine areas were clipped back to 4 cm, the normal height of the surrounding vegetation subjected to continued natural grazing; those in short-turf areas were clipped to 2 cm. Preliminary results (Table 7.15) indicate that, on improved grasslands at least, the current pressure of grazing reduces vegetational productivity to a level considerably below that which it would achieve if ungrazed.

If total potential production of an ungrazed sward — one that has *never* been grazed — is assumed to be closest to that of a latrine sward (normally grazed relatively lightly, by cattle alone) exclosed for a full two-month period, then yields of other grazing regimes may be compared against this as a base-line. (In fact such a figure is probably an underestimate of total ungrazed production, so conclusions are conservative.) Heavy grazing, of 'pony-type' (i.e. cropped every two weeks to a height of 2 cm), reduces production to 0.45 of this 'maximum' yield (Table 7.15). Even if relationship to 'latrine swards' is not strictly legitimate (in that

Table 7.15: Effects of different levels of artificial grazing on grassland forage yield. (Values are total yields in g/m^2 obtained over 16-week period in 1979)

Grazing regime	Frequency of clipping		
	2 weeks	4 weeks	8 weeks
Short-grass areas (cropped close by ponies)	204	240	270
Latrine areas (grazed as by cattle)	324	372	456

animal grazing has been so heavy for so long on these Forest grasslands that the basic soil conditions of latrine and pony-grazed areas are not truly equivalent), so that analysis is restricted to comparisons 'within type', production can be seen to fall — in both types of swards — as grazing pressure increases (yield for swards clipped fortnightly being about 70% of the yield of the same swards clipped at two-month intervals). All levels of grazing *reduce* yield. There is no evidence that on these Forest grasslands light grazing stimulates productivity; alternatively, all levels of grazing intensity imposed are above this 'stimulation' threshold.

Chapter 8

The Effects of Grazing on the Forest's Other Animals

Grazing and the Forest's Smaller Herbivores

As with the larger herbivores, the smaller animals of the Forest must find in their environment food and shelter if they are to survive. Different species differ in their food habits and in the type and degree of cover they prefer, but distribution of each species is restricted to those habitats which provide their particular requirements in food source and vegetative cover.

The various common British species of small rodents, for example, Wood mouse (*Apodemus sylvaticus*), yellow-necked mouse (*A. flavicollis*), field vole (*Microtus agrestis*) and bank vole (*Clethrionomys glareolus*), have all been shown to exhibit predictable variation in relative numbers — even in whether they occur at all — in relation to the available cover in a habitat. Thus, the field vole is the only species which prefers to live in open environments, unshaded communities of uniform and short ground vegetation; bank voles clearly prefer dense cover, and tend to be characteristic of hedgerows or woodland edge, habitats where both ground flora and shrub layer are at their most dense. Neither wood mice nor yellow-necked mice appear to display any particular preference for cover (Hoffmeyer 1973, Corke 1974) and, particularly in areas where population densities are high, may be found in a variety of different habitats. Thus, the wood mouse, while essentially a woodland species, is commonly found in 'bank vole habitat', or even open fields, when competition there is low. Although intrinsic *preferences* are thus less clear-cut than those of field voles or bank voles, the mice do not survive equally well in all types of habitat, and in the face of competition from other species may be seen to be more restricted in distribution. Hoffmeyer (1973) showed that the wood mouse restricts itself far more

165

tightly to its optimal habitat in the presence of the more aggressive yellow-neck; in the same way, while wood mice may colonise 'bank vole habitat' where voles are rare, they avoid the same habitats when bank vole populations are more abundant (Corke 1974).

Habitat preferences are not determined just by vegetative cover. While we have noted that the yellow-necked mouse shows no apparent preference for any particular degree of cover (Hoffmeyer 1973, Corke 1974), it is none the less found predominantly in woodlands (even in the absence of competitors). Further, throughout most of Europe it is most abundant in woodlands with a high proportion of beech or hazel (Schröpfer 1983). In autumn and winter, *A. flavicollis* feeds extensively on seed crops: beech mast and hazel nuts are staple foods. In feeding trials with other types of seeds such as acorns, the animals were found to be unable to maintain body condition; Schröpfer suggests that their preference for beech or hazel woodlands is thus a direct response to food needs. In the same way, *Microtus*, as a grazer for preference, feeding on shoots and roots of green vegetation, is probably restricted to open habitats not because it actively chooses to live without any cover from weather or predators, but because only in such environments can it find adequate food. *Clethrionomys* (grazer and fruit eater) chooses dense vegetation for cover, and hedgerows or woodland edge because of high food density, and the reason why the wood mice can afford to be unfussy in their choice of habitat is probably that they are catholic feeders, taking shoots, seeds, even insect food, as it is available.

Within the New Forest, the availability of cover and food is determined not just by the availability of particular vegetative associations or habitats, but is directly limited by grazing. Grazing by cattle, ponies and deer has a dramatic effect on the structure of particular communities, affecting the degree of cover offered — in canopy closure, extent of shrub layer or ground cover — as well as availability of food. How does this affect the smaller mammals, which, we have just discovered, appear to have quite clear and specific habitat requirements? What are the rodents of the Forest and how do their diversity and abundance compare with what we might expect in similar areas not subject to such heavy grazing? From 1982 to 1985, Steve Hill surveyed the rodent populations of the Forest woodlands, heathlands and open grasslands, comparing numbers and diversity with those obtained on equivalent vegetation types just outside the Forest boundary. Results were striking: total density of small mammals within the Forest was far, far lower than in adjacent ungrazed areas, though this was due not to a general lowering of population numbers across the board, but rather to a reduction in the number of species. Somewhat naively we had assumed that, if heavy grazing reduced both cover and food availability in all vegetation types, then certain species, whose requirements were highly specific, would indeed be excluded, but that those which 'hung on' would be surviving in suboptimal habitats and would thus show poor performance in comparison to populations in habitats more closely approximating to optimum conditions. What emerged from Hill's studies was that the diversity of species within the Forest was indeed much lower than in equivalent areas elsewhere: whole species were missing from the Forest woodlands and heathlands which occurred in abundance outside. Yet those species which *did* occur rather unexpectedly performed just as well inside the Forest as outside.

Heathlands and grasslands are obviously profoundly affected by grazing. The physical structure of these open vegetation types displays marked contrast with heathland or acid-grassland sites not subject to such heavy grazing (Chapter 7, Plate 8). On heathlands, the only animals caught in 1,200 trap-nights were two wood mice; adjacent heaths beyond the Forest boundary hold large permanent populations of wood mice and large populations of harvest mice (*Micromys minutus*). The story is very similar within acid-grassland communities: wood mice were more abundant, but again were found to use the area irregularly, and even then only in the summer, when bracken grows dense in these areas and offers some cover; only three field voles (*Microtus agrestis*) were caught during the complete grassland survey, and all of these were in a single area where bracken and litter cover is unusually thick all year round. The absence of harvest mice and field voles in trapping records does not by itself mean that the species are absent from the Forest: Hirons (1984) has found remains of both species in pellets cast by kestrels feeding within the Forest. Densities are, however, obviously *extremely* low; even those individuals caught by kestrels may come from the few ungrazed areas that remain within the Forest, such as railway embankments, fields and fenced roadside verges (see below, page 175).

Within the Forest *woodlands*, diversity of small mammals is again low. Most woodland sites had resident populations of wood mice, but bank voles (*Clethrionomys*) were almost totally absent from all sites surveyed and only one site held a permanent population of yellow-necked mice (*Apodemus flavicollis*). Bank voles were abundant in woodland areas outside the Forest, and the fact that they reappear within the Forest when grazing is prevented makes it clear that their absence from Forest woodlands is not just geographical accident. As part of an experiment to investigate the secondary succession within Forest woodlands after relief from grazing, two 11-acre (4.5-ha) plots of mixed deciduous woodland were fenced in 1963. All grazing herbivores were excluded from one pen; in the other, fallow deer were maintained at a constant density (Chapter 7). Hill exploited this ongoing experiment, surveying mammal populations in the two, adjacent, pens. The ungrazed pen showed a mammal community identical to those established in woodlands outside the Forest, with healthy populations of both bank voles and wood mice; 50 m away in the grazed pen, bank voles were never recorded (Table 8.1).

All these changes in species composition and species occurrence of small rodents within the various habitats of the Forest are exactly as one might predict from knowledge of the habitat preferences in terms of cover and food for each of the species (pages 165-6). None the less, the results emphasise that the effects of heavy grazing on the ecological functioning of the Forest are not confined to its direct effects upon the vegetation.

In Hill's studies, populations of *Apodemus sylvaticus* appeared little affected by the heavy grazing pressure. True, the mice were largely restricted to woodlands and made little use of more open habitats, but, within the woodlands, wood mice maintained healthy resident populations. Populations seemed to be of similar density to those outside; numbers inside and out showed the patterns of seasonal fluctuation typical of most woodland rodents. Indeed, on whatever criteria differences were sought — in age or sex structure of populations, average weights, fecundity, patterns of survivorship — no such differences emerged (Hill 1985); the wood mice of the New Forest appear unaffected by the dramatic vegetational changes inflicted by grazers. This result in itself is fascinating, considering the dramatic effects on other rodent species, and it is perhaps significant that the one species that *does* survive in the Forest and appears to perform as normal is that

Table 8.1: The number of individuals of the two species of small rodent, the wood mouse (*Apodemus sylvaticus*) and bank vole (*Clethrionomys glareolus*), caught during ten days Longworth trapping within the two deciduous woodland Denny pens, June 1982 (from Hill 1985)

Species	Grazed pen	Ungrazed pen
Wood mice	8	7
Bank voles	0	15

single species which we might have selected from previous knowledge as an opportunist: unfussy about cover and catholic in its diet.

The Forest Predators

These changes in distribution and abundance of small mammals presumably will have repercussions on the various predators of the Forest who rely for all or part of their diet on rodent prey. A very preliminary analysis has been made of the winter diet of New Forest foxes (*Vulpes vulpes*) by Senior (unpublished), who analysed the contents of a number of stomachs obtained between October and February. The foxes obviously took vertebrate prey whenever it was encountered (and 28% of guts examined contained the remains of birds or mammals), but such prey items were rare and the foxes clearly relied heavily on invertebrate material. Almost all guts contained significant quantities of beetles or earthworms, as well as leaf litter which had presumably been ingested accidentally while foraging for invertebrate prey of this kind. Perhaps more surprisingly, most of the guts examined contained quantities of plant material which would appear to have been taken deliberately: between 10% and 15% of the food contents of most stomachs looked at consisted of fruit, fungi or other edible vegetable material. While it is tempting to suggest that such a diet is unusual for an acknowledged carnivore and that the heavy dependence on invertebrate and vegetable material is a direct consequence of the 'distorted ecology' of the New Forest system, such claims would be excessive. In fact, the diet revealed by Senior's analyses is not particularly unusual; and analyses of fox diet in more 'typical' rural environments, while they do perhaps show a lesser reliance on beetles and other insect prey, none the less show equivalent quantities of fruit and other plant materials taken and also mirror the relatively low contribution made by vertebrate prey items (Table 8.2).

Badgers (*Meles meles*) are specialist earthworm feeders. In a study of badger diet at other sites within Hampshire, Packham (1983) found that 100% of all scats collected in all months contained earthworm chaetae, and that earthworms almost always constituted a high proportion of the diet of individual animals. Badgers in the Forest remained earthworm specialists despite the fact that earthworm densities within the Forest woodlands and heathlands are exceptionally low (Table 8.3, Figure 8.1). As a result, badger populations are little affected by low rodent numbers. Their ecology *is*, however, markedly affected by the

Table 8.2: Preliminary data on the overwinter diet of New Forest foxes. Equivalent data for the diet of woodland foxes in Oxfordshire (from Macdonald 1980) are shown for comparison

| | Percentage frequency of occurence | |
	New Forest foxes	Oxfordshire foxes
Lagomorphs and small rodents	28	20
Birds	—	11
Insects	56	21
Earthworms (and soil or leaf litter)	70	60
Fruits	28	20
Fungi	42	—

Table 8.3: Densities of earthworms available to badgers in different habitats inside the New Forest and elsewhere in sourthern Hampshire (from Packham 1983)

| | Earthworm density: Mean numbers per m^2 (S.D. in brackets) | |
Habitat	New Forest	Itchen Valley
Pasture/lawn	150 (26.6)	192.2 (89.4)
Heathland	0	—
Deciduous woodland	18.4 (15.9)	116 (27.9)
Coniferous woodland	1.2 (1.9)	26.8 (15.1)
Mixed woodland	3.2 (3.7)	64.4 (31.1)
Rides and glades	45.2 (33.9)	—

low densities of the earthworms on which they specialise, which in itself may be an additional effect of the Forest's heavy grazing pressure.

In studies of the social structures of badgers in areas of Hampshire outside the New Forest, Packham demonstrated an apparently constant relationship between the number of badgers in a territorial group ('clan'), territory size and earthworm density, such that the number of badgers shows a constant relation to total number of earthworms within the range:

(Range area × earthworm density) ÷ number of badgers = Constant

A similar relationship has also been demonstrated for Scottish badgers

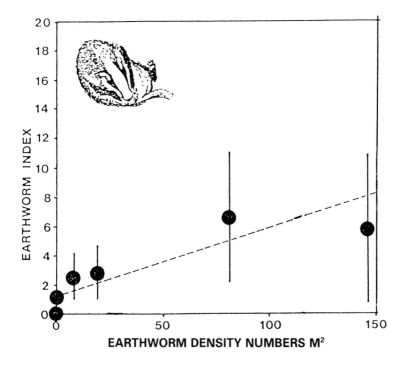

Figure 8.1: The relationship between relative earthworm abundance in the diet of New Forest badgers and density of earthworms in the habitat in which the diet was obtained (bars indicate one standard deviation around the mean)

Source: Packham (1983).

by Kruuk (1984). If Packham's 'formula' is extended to the New Forest, in calculation of territory size required to support a single badger, the low earthworm densities obtaining within the Forest result in predictions of a range so large that it would be impossible to defend. Linear dimensions increase only as the square root of area: as larger and larger range areas are predicted for clans of increasing numbers of badgers, so the length of the range boundary does not increase in the same proportion. But, even with 'theoretical clans' of eight or nine badgers, territory sizes predicted would have boundaries far too extensive to be effectively defended. Packham concluded that within the New Forest badgers should be more or less solitary, with extensive, overlapping home ranges which were not defended. Studies of the badgers themselves provided some empirical support for this: the badgers appeared to make no attempt to delimit range boundaries, and the usual marker latrines (concentrated around territorial limits in other areas) were not restricted to any clear perimeter lines; indeed, within the Forest, latrines appeared to be distributed randomly within the animal's range (Packham 1983).

In addition to its mammalian predators, the New Forest supports a remarkable diversity of birds of prey; populations of raptors may also be tremendously influenced by changes in the availability of rodent prey. Indeed, Colin Tubbs (1974, 1982) was the first to suggest that Forest predators in general might be influenced by grazing pressure and its effects on rodent numbers, drawing his conclusions from his now classic study of long-term changes in the populations of buzzards (*Buteo buteo*) within the New Forest. Tubbs noted that, in parts of England where rabbits are not readily available, buzzards appear to rely heavily on small rodents as prey; breeding success in these areas is directly related to abundance of such rodent prey. Although Tubbs had no direct data on abundance of small mammals within the New Forest, he none the less showed a clear correlation between reduced breeding success of the buzzards themselves and increasing grazing pressure from Common stock (Figure 8.2). Tubbs explains this relationship in terms of rodent numbers, suggesting that under high grazing pressure rodent populations are reduced, and buzzard breeding success declines accordingly: a not unreasonable presumption.

More recently, Hirons (1984) has examined diet and breeding success of other birds of prey of the New Forest whose diets would normally be expected to include high numbers of rodents. His studies of the diets of kestrels (*Falco tinnunculus*) and tawny owls (*Strix aluco*) produced surprising results (Hirons 1984). Small rodents contributed 66% by weight to the prey taken by tawny owls in the New Forest, comprising 42% of all prey items taken (Table 8.4). Studies of the diet of tawny owls elsewhere in Britain suggest that small rodents would normally constitute between 60% and 70% of prey items identified (Southern 1954, Hirons 1976). Owls in the New Forest thus seem to rely as heavily on small mammals here as they would elsewhere, despite reduced densities. In the tawny owl populations studied by Southern and Hirons, wood mice and bank voles each comprise about 30% of identifiable remains, with field voles contributing up to a further 13%. In the New Forest, voles are unavailable; but, within the Forest woodlands, *Apodemus* are present in normal densities. The owls respond to the reduced availability of voles by increasing their predation on wood mice and becoming wood mouse specialists. Thus, in the New Forest, as elsewhere, small

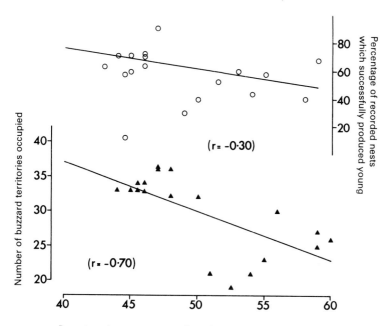

Figure 8.2: Relationship between grazing pressure on the New Forest and breeding success of buzzards. Data from Tubbs and Tubbs (1985). Note that, while success of nests, once a breeding attempt has been started, shows no significant effect from grazing pressure, the number of buzzard pairs attempting to breed declines as grazing pressure increases

rodents are the most important component of the diet of tawny owls, but this is achieved by taking very high numbers of wood mice: *Apodemus* accounted for 57% of the vertebrate prey items identified, while voles (*Clethrionomys* and *Microtus*) together constituted less than 10%.

Rodents still make up only 42% of the owls' diet; less in summer, when vertebrate prey constitutes only 33% of total intake. The balance is made up with a diversity of invertebrate food. Surprisingly, despite the importance of earthworms to tawny owls in other areas (where chaetae are found in nearly 100% of all pellets), *no* earthworm remains were identified from any pellets recovered from the New Forest (cf. badgers, page 170 above); it is clearly not these which compensate for reduced availability of vertebrate food. Dor beetles (*Geotrupes* and *Typhoeus*), however, contributed a surprisingly high proportion of the

Table 8.4: Analysis of 65 tawny owl pellets collected in the New Forest in 1982-3 (from Hirons 1984)

Prey	May-October			November-April			Overall		
	No. items	% items	wt	No. items	% items	wt	No. items	% items	wt
Wood mouse	26	35.1	45.0	21	27.3	55.4	47	31.1	49.1
Bank vole	2	2.7	3.1	6	7.8	14.1	8	5.3	7.4
Field vole	2	2.7	4.0	6	7.8	18.5	8	5.3	9.8
Grey squirrel	1	1.4	6.7	—	—	—	1	0.6	4.0
Rabbit	1	1.4	9.6	—	—	—	1	0.6	5.8
Common shrew	4	5.4	3.1	2	2.6	2.3	6	4.0	2.8
Birds	10	13.5	24.0	1	1.3	3.7	11	7.3	16.0
Slow worm	1	1.4	1.9	—	—	—	1	0.6	1.2
Coleoptera	27	36.5	26.0	41	53.2	6.6	68	45.0	3.9

prey of New Forest Tawny Owls, particularly in winter when they accounted for over 50% of items taken and formed 60% by weight of the total diet. The abundance of these beetles in the diet may reflect the large amounts of dung deposited in the New Forest by domestic herbivores and deer.

Field voles and bank voles have also been found to be an important element of the diet of kestrels (Ellis 1946; Davis 1960, 1975; Yalden and Warburton 1979). In the population studied by Yalden and Warburton, the two vole species together accounted for 73% of all vertebrate prey. In the New Forest, although bank vole remains are uncommon in kestrel pellets, the birds still feed extensively on field voles, which contributed 27% of identifiable vertebrate prey items recovered. Although the importance of *Microtus* in the diet is clearly lower in the New Forest than elsewhere in Hampshire (Hirons 1984) or in the studies mentioned above, the number of vole remains recovered from kestrel pellets is still higher than would be expected from Hill's independent surveys of rodent numbers: which suggested that field voles were completely absent from the closely grazed acid grasslands of woodland rides within the Forest. Most of the kestrel ranges from which pellets were collected in the Forest, however, contained roadside verges or railway embankments, which would provide habitat suitable to support at least small *Microtus* populations; in addition, many of the kestrels extended their range over adjoining farmland, which would provide an alternative source of voles as prey.

Overall, mammalian prey constitutes only 30.5% to the weight of vertebrate prey taken by New Forest kestrels (Table 8.5), as against the 73% estimated by Yalden and Warburton (1979). This shortfall is made good by a greater reliance on birds and by extensive predation on common lizards (*Lacerta vivipara*), which occurred in 45% of all pellets analysed (Table 8.5). Such heavy predation on lizards (which occurred in 59% of all pellets in April/May, when they spend much time basking in the open) appears to be a curiosity of the New Forest (in the study by Yalden and Warburton, only 12 lizards were recovered from the 219 pellets examined) and is probably a direct response to low densities of small mammals in open habitats within the Forest.

Tawny owl territories have been mapped in various areas within the Forest. Rather strikingly, there are no significant differences in average territory size between woodlands inside and those outside the Forest (Table 8.6), and territory sizes are similar to those found in deciduous woodland elsewhere (e.g. Southern 1969). Such a finding fits well with the previous discovery that the diet of New Forest owls is also essentially

Table 8.5: The frequency of occurence of different prey in kestrel pellets collected in the New Forest in 1982-3 (from Hirons 1984)

Period	April-May		June-July		Aug-Sept		Overall	
No. of pellets examined	104		57		29		190	
Prey	No. (%)		No. (%)		No. (%)		No. (%)	
Wood mouse	2	(1.9)	3	(5.3)	1	(3.4)	6	(3.2)
Bank vole	2	(1.9)	3	(5.3)	—		5	(2.6)
Field vole	28	(2.0)	16	(28.1)	1	(3.4)	45	(23.7)
Harvest mouse	19	(18.3)	1	(1.8)	5	(17.2)	25	(13.2)
Pygmy shrew	7	(6.7)	6	(10.5)	—		13	(6.8)
Common shrew	22	(21.1)	8	(14.0)	3	(10.3)	33	(17.4)
Small mammals							137[a]	(72.1)
Birds	27	(26.0)	8	(14.0)	2	(6.9)	37	(19.5)
Common lizard	61	(58.6)	17	(29.8)	8	(27.6)	86	(45.3)
Earthworms	1	(1.0)	—		—		1	(0.5)
Lepidoptera	—		—		24	(82.8)	24	(12.6)
Coleoptera	69	(66.3)	26	(45.6)	20	(69.0)	115	(60.5)

Note a. This figure includes unidentified mammals indicated by counts of upper incisors or fur.

Table 8.6: Mean territory size of tawny owls in different areas

Area	Territory size (ha)	No. of pairs
New Forest (Inclosure)	23.4 ± SE 1.9	11
New Forest (Open Forest)	17.5 ± 2.1	10
Outside Sides in Hampshire	17.9 ± 1.9	12
Wytham (Southern 1970)	18.2 ± 2.1	31
Forest of Ae (Hirons 1976)	46.1 ± 7.1	12

not dissimilar to that of tawny owls in general in that they are able to make up from 60% to 70% of their diet from small rodents, much as do owls elsewhere (page 173). Tawny owls are essentially woodland birds. We have already noted that New Forest woodlands support populations of wood mice of equivalent density to woodlands in areas less heavily grazed: it is thus not too surprising to find the owls behaving much as do owls elsewhere. But there is a difference. Hirons (1984) monitored breeding success of the owl populations through two years (1982 and 1983). In the early spring of 1982, rodents were scarce all over the country following a severe winter and a failed mast crop. Whether inside or outside the Forest, most owls did not attempt to breed at all, and no difference in breeding success could thus be detected between areas. In 1983, however, when breeding was attempted, the proportion of owl pairs successfully rearing young was much higher in ungrazed woodland outside the New Forest (65%) than within the Forest (25%).

Owls are woodland birds and, of all the Forest communities, woodlands showed the least difference in rodent densities from ungrazed areas outside the Forest; in more open habitats of the Forest, small rodents were almost totally absent. In 1982 the density of breeding kestrels was only one pair per 16 km^2; the density was lower still in 1983 (Hirons, pers. comm.). This compares with a figure of one pair per 4 km^2 in farmland outside the Forest.

Chapter 9

The New Forest — Present, Past and Future

The Present

Our studies over the past few years have shown how closely the various animals of the Forest community interact with one another and with their common vegetation, and how dominant are the effects of grazing at all levels of the system. Ponies, fallow, sika deer, and almost certainly roe — although as yet we have not the data to prove it — adjust their diet so that at any season they are feeding on those foodstuffs which offer maximum availability of nutrients, particularly digestible nitrogen (Chapters 4 and 6). Foraging behaviour is a major influence on patterns of habitat use: habitat selection (preference) and observed occupancy of different communities reflect the changing choice of foodstuffs and their changing availability (e.g. Chapter 3). Different vegetation types suffer markedly different pressure of use, and within communities seasonal changes in foraging are reflected in different levels of grazing pressure at different times (Chapter 7).

Animals' feeding behaviour and habitat use are affected not only by the vegetation. Through their own effects upon the vegetation they influence each other. Thus cattle and ponies, both preferential grazers, divide between them the different grassland types — and different zones within any common grassland (Chapter 3). Fallow, roe and sika deer show clear overlap in diet at certain times of year, and in seasons of food shortage potentially compete; throughout the year sika are clearly affected in terms of diet by interaction with ponies (Chapter 6).

The combined impact of all these larger herbivores upon the Forest's vegetation is colossal: the effects are dramatic (Chapter 7). Grazing acts to eliminate particularly favoured or sensitive species from a community. Species diversity falls in woodlands and open vegetation types

178

alike: grassland swards are remarkably species-poor, dominated by a few resistant species. Physical structure is altered: in woodlands, heathlands or grasslands there is little or no vegetation between 0.1 and 1.8 m — the whole structural layer is missing. Grazing prevents regeneration: the age structure of Forest trees reflects past periods of lesser and heavier grazing (Peterken and Tubbs 1965), and current grazing pressure *still* prevents regeneration within heavily used woodlands. Finally, productivity and nutrient status of the vegetation is affected. In theory, relatively light grazing may stimulate productivity (e.g. McNaughton 1979), while heavier grazing pressure will reduce vegetative production. In the New Forest, all vegetation types surveyed are so heavily grazed that productivity is suppressed: no sites showed increased production as a result of grazing (Chapter 7). At the same time, alteration to patterns of nutrient cycling within and between communities, combined with the fact that the greater part of a rather limited nutrient pool is bound up into animal tissue, result in continued impoverishment of soils subjected to heavy grazing.

It is perhaps appropriate to examine the more detailed results of our analyses in relation to comparable figures which are available from similar studies elsewhere. Bobek *et al.* (1979) estimated that red and roe deer in deciduous forest in southern Poland removed 46 kg/ha of browse material, some 28% of annual production; herbivore consumption of holly (*Ilex aquifolium*) in the New Forest, at approximately 7% (Table 7.4), is far lower and clearly has little immediate effect upon the holly itself. Offtake of deciduous browse species today, is low, but this should not be interpreted as of low impact: the reason why offtake is so small is that no understorey browse species other than holly persist within the Forest in any great abundance, having been eliminated by browsing pressure in the past. Impact of grazing on New Forest grasslands is colossal, with between 70% and 130% of growing-season production removed by grazing herbivores almost as soon as it is produced, far in excess of that recorded in the literature for almost any other *natural* grazing system. The open grasslands of the East African plains support a diversity and abundance of grazing ungulates perhaps unrivalled anywhere in the world, yet equivalent estimates for offtake (Wiegert and Evans 1967) suggest that from 30% to 60% of primary production only is removed.

This tremendous pressure on New Forest grasslands is of course a function of the distinct habitat preferences shown by the herbivores and the extremely uneven pattern of grazing pressure that results. Thus, some 50% of all animal grazing time is concentrated on grasslands which together comprise only 7% of the available Open Forest area (page 63), while other communities such as heathlands, despite their enormous acreage within the Forest, sustain but little grazing pressure (less than 9% overall of the pressure taken per m^2 of grassland). The net effect on the grasslands is a dramatic decline in productivity. Typical daily offtake for the improved grasslands of the Forest during the main growing season (May-July) is calculated as 3 g/m^2 per day, and conservative estimates suggest that over-grazing suppresses production of these grasslands to less than half of their assumed 'ungrazed' yield.

An increase in grazing intensity on sheep pastures in Australia (Vickery 1972) resulted in an increase in production up to stocking rates of 20 sheep per hectare. Production was least at ten sheep per hectare, and declined again with stocking densities higher than 20. A similar pattern of initial increase in production followed by gradual decline is well documented by other authors. McNaughton (1979a) presents a detailed analysis of the relationship between grazing and above-ground primary production for the Serengeti. It is notable that in the wet season (= main growing season) production increases only slightly with light grazing, up to offtake levels of 5 g/m^2 per day.[1] Stimulatory effects of light grazing were, however, never demonstrated by our studies of New Forest grasslands, and all levels of grazing observed by us reduced yield.

It is clear that patterns of dunging and urination also affect vegetational dynamics. Crawley (1983) quotes Harper in noting 'that to be covered by a cow pat is disaster for most plants; if they are not killed by the darkness or by the concentrated nutrients, they may have their living space usurped by creeping plants which grow over the top of the dung' (and others whose seeds may have been deposited in the dung itself) (Harper 1977). A cow pat has a long-lasting impact upon the vegetation, both because fouled areas are avoided and thus released from grazing and because a different community of plants establishes in the nutrient-rich conditions. But more significant perhaps is the effect that herbivore *behaviour* may have on nutrient distribution over the area as a whole: by grazing in some areas and defaecating in others, the herbivore may cause major discontinuities and translocation of nutrients (pages 6-7). Within the New Forest, clear redistribution of nutrients occurs on the Forest grasslands: the grazing and eliminative patterns of ponies and cattle result in a complex mosaic of patches

---

[1] Although this is higher than the figure quoted for offtake from New Forest grasslands (3 g/m^2 per day), values for forage production were also *extremely* high: consistently above 20 g/m^2 per day (McNaughton 1979a).

within the community. There are marked differences in the nutrient status of pony-grazing areas and cattle-grazing areas (= pony latrines), with continued loss, particularly of potassium and phosphorus, from the main pony-grazing areas (page 162).

These differences in nutrient status *and* differences in grazing pressure and 'style' cause noticeable differences in the species composition of these areas as well (Table 7.14). Other changes in species composition can be noted in grasslands and woodlands by comparison with equivalent sites beyond the Forest boundary and subjected to lower grazing pressure, or by comparison with historical records of past species structure. The main changes are recorded in detail elsewhere (Chapter 7) and there is no need to repeat them here. It is clear, however, that the changes in species composition and diversity of the different Forest communities are due to (i) altered nutrient status, (ii) grazing elimination of preferred or particularly sensitive forage species, and (iii) relief of competitive pressure on more graze-resistant or graze-tolerant species. Within woodlands, loss of more palatable species such as willow, hawthorn, blackthorn and birch shows clear parallels with the effects of grazing by cattle and sheep in deciduous forests in Europe (e.g. Adams 1975). In the patchy flora of the woodland floor, grasses increased in both abundance and diversity where grazed (Denny pens experiment, page 156). Similar results were shown by Spence and Angus (1971) in an analysis of the effects of large herbivores on *Combretum-Terminalia* woodland in Uganda: in plots protected from grazing, palatable species of shrubs and trees, such as *Desmodium* and *Indigofera* spp., increased from an abundance of perhaps 40 plants per hectare to nearly 654 per hectare; in the same plots, abundance of grasses such as *Sporobolus pyramidalis*, *Hypparhenia filipendula* and *H. rufa* declined when grazing pressure was removed.

Heavy grazing of New Forest grasslands causes a marked decline in diversity and a clear shift in species composition. Sensitive species such as clovers (*Trifolium pratense* and *T. repens*) are eliminated from the sward; there is a dramatic increase in the dominance of resistant grass species (notably *Agrostis tenuis*) and more prostrate, rosette-shaped forbs, which by their growth form are better able to escape severe defoliation (Table 7.13). Exactly comparable changes are recorded for grasslands in the Seregenti by McNaughton (1979), who notes that the shift in species composition between grazed and ungrazed areas is the result of replacement of tall, stemmy species, abundant in areas where grazing is prevented, by prostrate, low-growing species resistant to grazing. McNaughton suggests more generally that there is an adaptive trade-off among the species between ability to sustain sufficient leaf area and photosynthetic potential under intense grazing (facilitated by a prostrate growth form) and upward growth required in competing with other plants for light and space (facilitated by investment in stem production). In the Serengeti system studied by McNaughton, species like the *Pennisetum*, with a heavy investment in stems, are very susceptible to damage by grazing, but win at light competition when grazing is reduced (page 5); by contrast, prostrate species like *Andropogon greenwayi*, quickly out-competed in the battle for light in ungrazed swards, dominate the community in areas more heavily grazed.

Within the more immediate context of the New Forest, it is particularly interesting to consider the fate of those lawns re-seeded in the 1950s and 1960s with results of studies by Jones (1933) on the long-term effects of grazing sheep upon sown pastures of different compositions at Jealott's Hill in southeast England. These experiments are peculiarly relevant because they demonstrate so clearly the effects of different grazing patterns: pastures sown with exactly the same seed mix and thereafter managed according to exactly the same cultural practice were simply subjected to different grazing regimes.

In one trial where a rye-grass/clover (*Lolium perenne/Trifolium repens*) mixture was sown, changing only the timing of grazing produced an almost pure rye-grass sward in one case (6% clover) and a predominance of clover (62%) in the other. This was brought about by grazing regimes which tipped the competitive balance in favour of the grass in the first case and of the prostrate-growing clover in the second. To produce a rye-grass sward, the animals were kept off the pasture in early spring so that the rye-grass tillers produced vigorous root and shoot systems. When the sheep were introduced to the pasture, the grass plants had substantial reserves of root and shoot and were able to recover rapidly from defoliation. Despite the attentions of the sheep, they were able to over-top and out-compete the clover plants. The clover sward was created by introducing the sheep to the grassland very early in the year (during March and April), so that the first rye-grass leaves were grazed off as they developed. The sheep were removed before they grazed down to the lower clover plants. By the peak growth period in May and June the clover plants were in fine condition, but the rye-grass had poorly developed root and shoot systems, and suffered from the combined effects of defoliation and competition from the clover (Jones 1933). Clearly, management has a profound effect on the results of grazing: the New Forest grasslands perhaps represent a third permutation of Jones's experiments, one where grazers were permitted early in the year but not 'removed before they grazed down to the lower clover plants'!

In the long term, heavy grazing at this level can affect successional processes, and ultimately result in vegetation loss and desertification. Certainly within the Forest, current and past levels of grazing have been sufficient to prevent significant regeneration of woodland; yet other communities, though somewhat 'distorted' by the continuing pressure from herbivory, seem none the less relatively stable.

The Forest — Past

All the effects of grazing upon vegetation described are markedly influenced by *intensity* of grazing. Over its 900-year history, the Forest has seen marked fluctuations in the numbers and type of grazing stock (Table 4.1); such changes in grazing pressure will have meant differences in 'immediate' ecological function of the Forest system at different times — and also leave a legacy for the future. The New Forest today is affected strongly by current grazing pressure, but is also markedly influenced by its past.

As noted in Chapter 2, deer, and particularly fallow deer, were a significant and probably fairly constant pressure on the Forest from its establishment in the eleventh century until the Deer Removal Act of 1857, perhaps having been present in numbers up to 8,000 or 9,000. From then numbers suddenly declined, and have only recently regained their current levels of about 2,500. Changes in the numbers of Common stock grazing on the Forest are harder to determine. From the eleventh to the eighteenth centuries there must have been considerable populations, but no formal records of animal numbers exist prior to 1789. Since that time there have been dramatic fluctuations in overall numbers as economic climates have changed; and, more subtly, there have also been changes in the relative balance of the two main species depastured, cattle and ponies.

Between 1789 and 1858, marking fees were paid for between 2,000 and 3,000 ponies, but for up to 6,000 cattle. In the latter part of the nineteenth century, numbers of cattle and ponies depastured were more nearly equal (between 2,000 and 3,000 head). From then onwards numbers of cattle have continued to decline; numbers of ponies have increased steadily until the late 1970s (with exceptional decreases of all stock during the war years). Thus, there has been a shift in commonable stock from cattle — bulk-feeding ruminants, which lack the ability to crop close — towards monogastric horses, which crop close, digest poorly and rely on a rapid throughput of forage. Among the cattle that remain, there has also been a shift from predominantly dairy herds towards beef and store cattle. Initially cattle turned out on the Forest grazings were predominantly dairy animals, with many Commoners turning out a few 'house' cows and heifers. More recently this practice has declined, while a few Commoners have started to run relatively larger herds of store cattle upon the Forest. Such herds are more likely to be kept upon the Open Forest throughout the year (with the aid of supplementary feed), and this may ultimately lead to a more significant effect upon the vegetation. Changes in the relative balance of cattle and ponies, or store and dairy cattle, closely reflect changes in market prices. At the same time they are influenced by a social change in commoning. Over the years we have witnessed a shift in use of Commons, from supporting a subsistence cottage economy to supporting a form of commercial livestock agriculture.

Such social and economic changes are clearly of tremendous significance in the Forest's ecology. Common stock *now* represent the most significant grazing influence upon the Forest and their changing fortunes, past and future, will clearly have a powerful effect upon the Forest.

Colin Tubbs has done more than any other to tease out the social history of commoning in the New Forest, painstakingly picking over hundreds of reports, notebooks, official and unofficial records, and piecing together the few scraps of information culled from so many sources to try and gain a picture of the changing fortunes of Common agriculture in relation to the changing ecology of the Forest itself (Tubbs 1965, 1968). His study is a masterpiece of patient historical detection, in researching the records and in interpretation; it seems hard that we may now draw on the results of all his labours and initiative so painlessly.

Although records before the initial Registers of Claims of 1635 and 1670 (page 23) offer only fragmentary evidence about commoning in the years leading up to the seventeenth century, it is clear from what evidence is available — presentments at Forest Courts in respects of abuses of Common Rights, or petitions from Commoners alleging unreasonable infringement of these rights — that the area supported a vigorous pastoral economy. Sheep, cattle and ponies were grazed upon the Forest from the thirteenth to sixteenth centuries. Some of the biggest graziers were the large landowners — the big religious houses at Beaulieu, Christchurch and Braemore — but at the same time it is clear that *all* Rights of Common were also extremely important to occupiers of smaller holdings: cottagers who could make a subsistence cottage economy considerably more comfortable by exercising their rights to fuel, animal bedding and fodder for their few livestock. With the dissolution of the monasteries in the middle of the seventeenth century, and the accompanying changes in land ownership, patterns of commoning also began to shift. In the Registers of Claims compiled in 1635 and 1670, most claims (whether entered individually, or as part of a 'block application' by a large estate on behalf of many tenants) were in respect of smaller holdings. The Register shows three main classes of people exercising Rights of Common over the Forest: Lords of the Manor with land within or adjacent to the Forest; his tenant farmers and copyholders; and a class of freeholders or yeomen with their own small farms. Most claims were in respect of cottagers with only a few acres; even the holdings of tenant and yeomen farmers were generally smaller than 50 acres (20 ha) and more usually between 15 and 30 acres (6-12 ha). Now, too, Commoners were exercising their rights not just in order to support a cottage economy, but to support agriculture; Tubbs notes that the majority of holdings were too small in themselves to be self-supporting, and we may deduce that particularly grazing rights were being used to support an economy of small livestock farmers, entirely reliant on the use of Common lands.

The picture remains essentially unchanged throughout the eighteenth and nineteenth centuries. The majority of claims submitted for Rights of Common were for freeholdings of less than 80 acres (30 ha). A number of claims were entered by large estates, but once again each of these submitted a general claim on behalf of numerous tenants who occupied holdings usually of less than 50 acres. Of claims allowed in 1858 for freeholders, 629 were for properties of less than 30 acres, and 400 of these for cottages with between 1 and 4 acres (0.4-1.6 ha). In 1944, Kenchington quoted the total number of holdings with Common Rights on the Forest as 1,995; of these, 731 were below 5 acres (2 ha) and only 396 were above 50 acres. Kenchington notes that even the tiny holdings were equipped not as smallholdings proper but as miniature farms. Two-thirds of the holdings represented the main source of livelihood for the occupiers. Clearly, the shift in use of Common Rights away from merely assisting a cottage economy towards supporting grazing agriculture on a large scale, a shift that we identified even in the seventeenth century, had continued still further. Evidence given before Select Committees in 1868 and 1875 noted that 'The Right of Common of grazing enabled a Commoner to maintain three times as many cattle as he would be able to without that right.'

Numbers of cattle and ponies depastured on the Forest today are as high as those of the peak years of the 1880s, but the actual number of Commoners exercising grazing rights is considerably smaller. In 1858 1,200 holdings were registered; today Rights of Common are exercised only by a few hundred. A survey published recently by the Countryside Commission (1984) allows more objective analysis of the social and economic context of commoning today. The picture is still one of many small farmers with small properties turning out small numbers of stock, rather than a Common dominated by a few landowners turning out huge herds. The majority of commoners (75-80%) turn out 20 head of stock overall, or less, onto the Forest grazings; 52% of all practising Commoners work properties of less than 20 acres, and only 8% own more than 50 acres (Tables 9.1, 9.2). The pattern of grazing has not changed — but the social context has. Only 10% of the holdings are now worked full-time, and, for 50% of Commoners surveyed, their 'agricultural' enterprise contributes less than 10% of total income. More fundamentally, only 44% of existing Commoners are agricultural or manual workers within the Forest; 24% are professional or clerical. Commoning, which was once an intricate part of the livelihood of the New Forest, 'has become basically a grazing option offering an interesting way of life and supplementary or seasonal pasture to a few small-holding Commoners' (Countryside Commission 1984).

In summary, it would appear that the socio-economic status of commoning has changed more markedly in the last few years than over seven centuries before that. From the thirteenth to sixteenth centuries, Common pasturage was exploited by two very distinct classes: by big landowners, for free-range pasturage of large herds of grazing stock; and, almost at the opposite extreme, by cottagers with a small plot of

Table 9.1: Sizes of herds pastured by today's Commoners (from Countryside Commission Report 1984)

	Commoners involved	
Size of herd put out	Number	Percentage of all Commoners
0-25	298	84
26-99	49	14
100+	6	2
Totals	353	100

Table 9.2: Size of holdings owned by present-day New Forest Commoners

Acreage	Percentage of Survey
< 1	8
1-5	21.3
6-10	12.0
11-15	5.3
16-20	5.3
21-30	2.7
31-50	9.3
51-100	5.3
> 100	2.7

land, as additional grazing for their few domestic animals. Through the seventeenth, eighteenth and nineteenth centuries, there was a gradual shift in the use made of the Forest Commons; Common Rights were used primarily to support small farming enterprises, allowing the owners of relatively small holdings to practise as economic livestock farmers on a scale far beyond what would be expected of their limited acreage, because of the extra grazing offered by the Open Forest. This, in part holds true today. Much of the property around the edge of the Forest, together with those pockets of agricultural land within the Forest boundary, consists of small or comparatively small holdings (mostly still less than 50 acres). Yet many of these are run as miniature farms, whose profitable management rests largely on exercise of their Common Rights. Yet things *have* changed: commoning is less central to the life of all but a few of today's Commoners, and the 'social' context of commoning is clearly far removed from that of earlier centuries.

The extent to which Common Rights other than those of pasturage have been exercised has also changed over the years, and the changes reflect in part the increasingly 'agricultural' use of Common grazing. Through the thirteenth to seventeenth centuries, every claim for recognition of Commons was for all rights — pasture, mast, estover, turbary and marl; where Commons were supporting subsistence economies, these other rights made a considerable difference in reducing the costs of running a home. But, as the relative balance of different 'types' of Commoner changed, with the majority using Common pasturage to support their holdings more and more as miniature farms, so the direct value and importance of the other Common Rights declined. Various other factors also reduced the actual economic value of such concessions. Successive Acts of Parliament restricted the quantity of fuel each commoner might take under Rights of turbary and estover; Rights to estover were bought out by the Office of Woods as deliberate policy in order to try to check unbridled exploitation of the Forest timber; and the importance of all concessions dwindled as social circumstances changed, the Forest was opened up and access improved by road and rail.

From the thirteenth to the seventeenth centuries all claims for Common Rights included claims for estover, turbary and marl; by the end of the nineteenth century, Rights of marl and estover were little exercised. In the 1858 Register of Claims 1,500 holdings claimed Right of turbary; by the turn of the century even these rights were rarely claimed. Turves are cut by fewer than a dozen holders of turbary rights in the Forest today, and Rights of estover and marl are exercised only sporadically (Table 9.3).

The Future?

Clearly, both the social context and the economics of commoning are changing, and fluctuations in the numbers of stock pastured on the Forest will continue unpredictably as markets rise and fall. But Commoners' animals will probably always remain a significant influence on the ecology of the Forest; indeed, the maintenance of some level of grazing is absolutely essential if we are to retain the peculiar vegetational structure of the Forest as we know it. In addition, commoning has more than a purely ecological relevance; it is in itself a curious social and historical phenomenon, worthy of conservation in its own right. So, what of the future? How can the Forest continue to offer all things to all men — continue as an economic silvicultural enterprise, support Common agriculture and maintain its unique ecological value? What should the balance of interest be, and how should it be maintained?

Table 9.3: Current use of Common Rights (from Countryside Commission Report 1984)

	Number
Pasture	73
Mast	23
Pasture (sheep)	5
Turbary (turf cutting)	9
Estovers (fuel wood)	8
Marl	4

It is frequently claimed that with its present densities of Common stock the Forest is currently 'heavily overgrazed'. Over-grazing is an emotive but rather nebulous concept. Do we mean that grazing pressure on the Forest is so high that animals are no longer able to maintain condition, or that from a vegetational point of view grazing is such that productivity is suppressed? Or do we mean that grazing pressure is causing a decline in the 'ecological value' of the area through reduction in diversity of species composition and structure, or that it is preventing woodland regeneration? According to which criterion we adopt, we may feel that certain areas of the Forest may or may not be considered overgrazed.

Equally importantly, different community types may differ in the degree of 'over-grazing' experienced. We have stressed already that grazing animals do not use all habitats equally and that some sustain much higher impact than others (Figure 7.2). Further, different patches of the same community type suffer different grazing pressure dependent on their juxtaposition with other communities or geographical position within the Forest. Thus, certain areas of the Forest or certain specific communities may indeed be overgrazed, while at the same time others could be claimed to be *under*grazed.

The maintenance of some level of Common grazing (with, in consequence, some degree of vegetational management aimed at supporting the grazing stock) is clearly a crucial factor, but what should that level of grazing be in order to create and maintain ecological diversity and not diminish it, and how far should vegetational management be directed at improving the grazing for Common stock?

From the data we have presented so far, it is possible to attempt to calculate a crude estimate of the Forest's capacity to support grazing livestock. Such calculation is blatantly agricultural in its approach and considers the capacity of the Forest's vegetational resources to support stock; such a stocking density may *not* be that which would be 'desirable' in terms of maintenance of ecological diversity. None the less figures for required forage offtake per day per individual cow or pony may be related, in theory, to measured forage productivity, in order to determine the potential stocking density of the Forest for Commoners' animals and other herbivores.

Of course, such a calculation is grossly oversimplified. As discussed elsewhere (de Bie 1982, van der Veen 1979), one cannot expect merely to divide total annual forage production of an area by required animal offtake and come up with a figure for environmental carrying capacity. Forage availability is not the only factor affecting animal density;

animal numbers may be limited by other factors such as availability of shelter, etc. Even within the forage component, forage availability — and thus carrying capacity — changes in different seasons. Nor is 100% of forage produced available to the animals; a proportion may be inedible or unpalatable, while even whole areas of potential forage may also be relatively under-used because they lie outside the animal's normal range. In any case, animals do not require just a 'total energy intake' irrespective of how it is derived, but may require different things from different foodstuffs, and may require a specific mixture in the diet (and are thus limited not by total forage availability, but by the availability of the most limited of these resources). Other major factors affecting carrying capacity are purely behavioural. Animal densities may be restricted because of considerations of range use and social organisation. Some of these factors may be taken into acount. Our data for the New Forest present information for seasonal changes in productivity, while the fact that 'not all vegetation types are alike' to the animals may be accounted for in that seasonal productivity is calculated separately within each vegetation type and may thus be related directly to figures for required offtake also derived for that particular vegetation type. With carrying capacity thus derived in any season in terms of that vegetation type offering minimum supply in relation to demand, full account is taken of the specific requirements of cattle and ponies from their different forage types in areas of mixed vegetation (Putman *et al.* 1981).

Calculations are presented here, by way of example, for New Forest ponies. For each two-monthly period of our vegetational measures we have a figure for the productivity of each community type, or for forage species within them. For the same two-monthly period we have a figure for amount of offtake required per animal *from that particular community type*, or *forage species* (calculated as offtake of that community type per grazing animal, or total intake from that community: data from Chapter 7). For each season, it is then possible to calculate *total usable productivity over the whole Forest* for each community type/forage species by multiplication of its productivity per unit area by known available area of that forage type in the Forest as a whole (Table 3.2). This can be divided by total offtake required per animal of that particular community/forage species in that season; the lowest figure (for productivity/requirements of animals from that particular community) for whatever community must represent the maximum numbers of animals the Forest can support at that season, even if other communities show surplus productivity (Table 9.4).

Table 9.4: Carrying-capacity calculations: total production of each vegetation type over the Forest as a whole is divided by required offtake of the same vegetation type per pony to derive an estimate of the potential number of animals supported (see text, page 190)

Month	Re-seeded lawn	Streamside lawn	Commoners' improved grassland	Acid grassland	Woodland glades
January	—	113	139	182	—
February	? 47	424	423	392	—
March	1,228	164	—	—	—
April	1,310	493	—	—	—
May	594	218	376	290	116
June	832	384	622	1,812	344
July	1,510	556	770	1,503	133
August	892	450	244	510	8
September	459	198	280	430	—
October	152	172	66	434	—
November	—	186	—	—	—
December	—	203	—	—	—

(While this calculation does take into account seasonal changes in forage productivity and offtake and also allows for differing importance of different forage types, we are to use in our calculations total vegetational area of a particular community calculated for the Forest as a whole. We are not able to take into account the fact that not all areas of the same vegetation type are used equally intensively, that some are perhaps used hardly at all owing to the fact that they do not fall within the home range of many animals. Nor can we take into account the effects of social interaction; or the fact that not all *apparently* available forage may be acceptable or palatable. Our calculations thus assume that all forage produced within the Open Forest is (a) available to the animals and (b) equally available to them. With these reservations, then, we may still view the calculation as some form of approximation.)

Before we even examine the results of such calculations, certain facts are immediately apparent from raw figures for vegetational production and offtake. Available heathland, both wet and dry heath, far exceeds the extent of animal use of heather or other heathland plants. Even on wet heath, which becomes an important community in midsummer, when grass is in short supply, production consistently exceeds offtake. Similarly acid grasslands, woodland glades and woody browse are all in superabundance within the Forest. By contrast, there is severe competition for the favoured communities and forage types. Demand for brambles and gorse exceeds supply in certain areas in appropriate seasons, while offtake from all 'sweet' grassland communities (RL, SL, C1) consistently equals or exceeds production, even to the extent that in these communities productivity is depressed. These, then, are the vegetation types limiting the densities of animals upon the Forest.

Examination of Table 9.4 would suggest that, in fact, acid-grassland communities and woodland glades may also be potentially limiting after all; we must, however, remember that carrying capacity was estimated by relation of offtake to *production*, rather than production + standing crop (because measured standing crop in any one month is so close a function of *current* grazing pressure). We know from other elements of the study that in both woodland glades and acid grassland there is a high standing crop the year round: carrying capacity is therefore probably considerably higher than might be suggested by considerations of production alone.

Where true availability *is* more closely related to forage production (RL, CI, SL) it is clear that resources are indeed limited, as we have concluded from other evidence. No one of these communities, however, is likely to be limiting in its own right: the animals are well able to use them as alternatives. Thus, *if* we assume from these, and other, considerations that grasslands are likely to be limiting to stocking density, and *if* we assume that our calculations here are reasonable, we might derive a final, seasonal figure for carrying capacity as the sum of the separate capacities of RL, SL, CI (Table 9.5).

Such a calculation is fraught with assumption, and compounds many sources of error. Moreover, any conclusions drawn from it *assume* that the animals will continue to display in the future, and under altered conditions, the same pattern of community use as they do at present, for the calculations are based on current patterns of community use. The calculation, while of interest, is thus *highly* speculative and its conclusions should be interpreted with caution!

Belief that the Forest is currently overstocked (in relation to Commoners' agricultural interests, and thus in the sense of its capacity to support stock) has repeatedly led the Commoners to press for management aimed at increasing the available grazing on the Forest. The grazing improvements traditionally sought include clearance of scrub and gorse from grasslands, and draining of bogs and streamside lawns. Such measures clearly run counter to conservation interests within the Forest, threatening structural diversity, threatening the integrity of whole communities, and if carried through would perhaps shift the delicate balance of conflicting interests in the Forest too heavily towards commoning.

In the past, there was a real need for Common Rights — and a proper use of those rights — in sustaining the livelihood of the cottagers and smallholders of the Forest. It was appropriate at that time that the

Table 9.5: Final estimates of carrying capacity for ponies

Month	Stocking capacity
January	—
February	c 900
March	1,800
April	2,200
May	? 1,200
June	1,840
July	2,840
August	1,600
September	940
October	400
November December	uncertain: 200-300

interests of commoning should be upheld and that those interests should be taken into account in structuring management policy. Now perhaps, with the new social role of commoning, there is not the same justification if commoning conflicts with other interests. But, in any case, the measures suggested could in practice prove counter-productive even to the Commoners themselves. We have noted that both cattle and ponies are both socially and ecologically so 'hefted' to the Forest grasslands that they are reluctant to leave them even when other communities offer more food. This remains true in autumn and winter, when animals remain on the exposed Forest lawns even in appalling weather. Gorse brakes and scrub offer some degree of shelter in such areas — and may be crucial in preventing excessive heat loss in the worst of winter weather. The effects of draining bogs and streamside lawns are equally dubious. The beneficial effects of periodic flooding, which flushes the ground with nutrients, make streamside lawns among the most productive communities in the Forest; they also begin to offer fresh grazing slightly earlier in the season than the 'dry' grasslands. The precise changes in vegetation following drainage of valley bogs are more difficult to predict because of wide variation in wetland conditions. Nutrient-poor bogs dominated by purple moor-grass (*Molinia caerulea*) and cottongrass (*Eriophorum* spp.) growing on oligotrophic peat are likely to develop an extremely short and scanty turf composed largely of *Carex* species if drained and heavily grazed. It is doubtful whether such vegetation represents an improvement in food resources, since the total yield of such areas is probably lower than in the bogs, where *Molinia* grows luxuriantly. Further, the effect of both drainage proposals is to increase the availability of grassland. By reducing the diversity of community type, asynchrony of production is also diminished. All 'drier' grassland types show peak production from May to July: at a time when in fact there is alternative forage in abundance on the wetter heathlands. Drainage of bogs and streamsides merely increases vegetational production in the only season of relative abundance: useful supplies of winter forage provided by *Juncus* spp. and the early bite provided by streamside turf or the *Juncus* and *Molinia* of bog communities will be reduced.

Such comments are somewhat conjectural. We can be more objective about the likely effects of other potential forms of direct vegetational management. For the most part, we have considered the entire herbivore-vegetation complex in these pages as if it were a completely natural system. Yet many, if not most, of the vegetational communities are already man-managed, or at least susceptible to such management. Effects of the various management practices adopted in the past may be evaluated — in terms of their effects on the vegetation itself *and* on the various grazing herbivores — and used to predict the likely response to future management.

Improved grasslands of various types prove among the most attractive grazing areas for both cattle and ponies, and indeed for the fallow deer of the Open Forest. As a result, they are somewhat over-utilised: almost all production is used immediately, and both productivity and nutrient quality of the areas have been significantly reduced by the continuous heavy grazing pressure. The animals are offered an increasingly impoverished food supply; and, in response to both heavy grazing and poor soil nutrients, botanical diversity of the areas is minimal (Table 7.13). The original seed mixtures used for re-seeding contained several species of herbage plants of which perennial rye-grass and wild white clover were usually the most important (page 43). Under conditions of continuous heavy grazing, these valuable herbage species were quickly eliminated and replaced by various less desirable native species, of which *Agrostis tenuis* is predominant. (It is worth noting, however, that the most successful of the sown species was red fescue cultivar: S 59, and where this was sown it retains a cover of around 22%.) Many of the problems outlined above are a direct result of hard, continuous grazing. If access to re-seeded areas were controlled, it would be possible to manage these pastures to achieve very substantially increased yields. It would also be possible to alter the balance of plant species, favouring more productive and nutritious species such as perennial rye-grass and white clover (see also page 198).

The improvement of several acid-grassland sites in the early 1970s through swiping of bracken and liming has probably been the most cost-effective of all grassland management. The removal of tall bracken has opened these areas to grazing animals during the summer months, and many of them are now heavily used by livestock. The applications of lime raised the soil pH significantly (from about 4.5 to about 5.7), and was probably the cause of a marked reduction in the highly unpalatable bristle bent grass *Agrostis setacea*. The higher pH, coupled with closer grazing, may also explain the increased plant-species diversity of these sites compared with unaltered acid grassland.

The yield of herbage on the improved acid grassland is as high as, if not a little higher than, that of most re-seeded lawns. There is, however, a possibility that with increased use they will experience the same depletion of nutrients that has occurred on re-seeded areas as a result of the grazing and dunging patterns of ponies. It will be interesting to see whether they maintain their present levels of productivity over the next few decades.

The other notable vegetation types markedly affected by management are the dry heathlands of the Open Forest, which are burnt or cut on a regular rotation (page 45). Our studies demonstrate that only a small proportion of heather is taken by large herbivores, and that ponies in particular take very little. Most heather browsing occurs in the immediate vicinity of lawns and roadside verges. Burning or cutting of heather does, however, significantly promote the growth of *Molinia* for a few years, and this provides a grazing resource which is particularly valuable at the beginning of summer before the grasslands are fully productive.

Most of these options for management of the vegetation are aimed at maintaining densities of grazing herbivores at current high levels. Many of the measures would conflict directly with conservation or other interests; more fundamentally, it is arguable that current stocking levels of herbivores are already too high. It is apparent that, at least in some habitats and in some areas, grazing really is excessive — by *whatever* criteria.

Control of animal densities is, however, difficult. The Forestry Commission attempts to maintain numbers of the various deer species at appropriate levels, by careful census and suitable culling; the Verderers, too, have the right to restrict the numbers of Common stock depastured. Clearly, the most obvious management possibility is to restrict *overall* the numbers of animals pastured on the Forest. Restrictions could be absolute (set by considerations of carrying capacity at that time when it is at its minimum), or more flexible (different densities at different seasons, or in different geographical areas of the Forest so far as any of these are relatively defined entities). Such restrictions, however, would be difficult to impose and still more difficult to administer. In addition, we can offer here no clear idea of what may be the carrying capacity of the Forest.

The alternative strategy is to alter in some way the *distribution*, both geographically and between different community types, of existing numbers of animals on the Forest. The two most striking features about the Forest grazing system are, as already noted, that

(i) Whole areas of the Forest are relatively less used than others. Even within one community type which may be quite regularly exploited by the Common animals, one 'patch' may be far more extensively used than another.

(ii) Many particular communities of the Forest suffer considerable pressure — to the extent perhaps of over-grazing — while adjacent communities, containing perfectly palatable forage species on which the animals are quite happy to feed, are left virtually ungrazed.

The reasons for these two phenomena are, we believe, purely behavioural; in addition, we think that it may prove possible to manipulate the animals, using these natural behavioural responses, in order to achieve a more even distribution.

Clearly, the problem of differential density of animals over the Forest as a whole could be solved purely mechanically, in containing animals within certain geographical areas by fencelines and road grids. This, however, would be politically contentious, and expensive, and would spoil the open nature of the Forest. In addition, it might 'penalise' animals restricted to currently less favoured areas. We believe that the relative over-use and under-use of different parts of the Forest is a result of home-range patterns, and that an alternative solution may be adopted in exploitation of this knowledge. As already noted in this volume (Chapter 3), both cattle and ponies are preferential grazers, and base their home range upon a focal area of grassland. With a restricted number of such grassland areas available, animals congregate on those areas containing such a lawn, with resultant uneven distribution over the Forest as a whole. Cattle often make protracted route marches to and from this focal lawn and thus may range quite widely from it. By contrast, ponies feed and move less widely and tend to remain on those communities in the immediate vicinity of the focal lawn. The distribution of ponies is markedly aggregated, and ranges cluster around these few favoured lawns. Large areas of perfectly palatable forage remain unexploited, because they are too far from a suitable grassland, and no range includes them.

As noted, this is more of a problem with regard to ponies than to cattle. But we suggest that a more *even* level of grazing within given communities over the Forest as a whole could be achieved by the artificial production of small areas of new grassland in appropriate regions of the Forest — to open up areas of acid grassland and heathland at present under-exploited. These grasslands need *not* be particularly large (approx. 10 ha) and would thus pose no threat to conservation of the communities into which they are introduced; but, by evening up grazing pressure on the Forest as a whole, they could in fact benefit conservation by relieving grazing pressure on currently hard-pressed areas.

This preference for lawns also lies behind the second problem, that of the differential grazing pressure between *different* communities within a home range; and lawns are very decidedly a two-edged weapon. As soon as grass production starts on the lawns, the animals forsake other feeding communities and move onto these grasslands. Instantaneous offtake virtually equals production, so that in practice there is almost at once very little grazing to be had. Once on the lawns the animals are loath to leave, even though abundant, palatable and nutritious forage is available in adjacent communities. Both cattle and ponies may move off the lawns to forage on wet heath, bogs or even dry heath around the lawn edge, but they will not willingly penetrate deep into these communities and soon return to the focal lawns. Huge congregations of animals thus accumulate on lawns where there is little to eat, while in the surrounding communities forage resources are under-exploited. So strong is this 'hefting' that it persists even long after the growing season is over: ponies and cattle are still on the lawns in large numbers in late autumn and early winter, and only gradually drift away. Here, again, the lawn hefting results in under-exploitation of alternative forages but over-exploitation of the grasslands themselves. As a result of all this, the lawns become severely overgrazed — to a degree where productivity is suppressed through pure lack of photosynthetic tissue and through continued nutrient impoverishment. The animals, too, are suffering: the concentrations of animals on the food-sparse lawns result in a forage intake by each pony or cow far below that which it could achieve on other communities. The problem is further exacerbated in the winter months, when lawn-hefted animals (or animals attracted to any exposed community for hay feeding) stand about in open habitats, with nothing at all to eat, while fully exposed to wind and cold.

Winter loss of condition could be decreased, forage intake the year round increased, and differential grazing pressure of the various forage communities *and* the impoverishment of the re-seeded and other improved lawns could, we believe, be simply resolved, by allowing only seasonal access to these grasslands. We suggest that, were these grasslands fenced and common animals allowed onto them only in late summer (from perhaps August till October/November), a number of benefits would result:

1. By resting the lawns for eight months of the year, or by allowing only intermittent grazing on them during this time, they would be permitted time to recover and the continued impoverishment would be arrested.

2. By preventing over-grazing during the growing season, greater

total forage productivity could be achieved for the time when the lawns *were* opened for grazing.

3. Shutting off the lawns during the summer would in no way deprive the animals of forage, since the grassland growing season coincides with the peak growing season of many other communities currently under-exploited. Indeed, by getting the animals off the lawns and into these other communities, forage intake would actually be increased.

4. More even grazing of all communities would result.

In addition, we believe that a further benefit might be obtained: better overwintering performance of stock left on the Forest.

More subtle management of this kind might help to restore the uneasy balance between grazers and the vegetation, might relax over-intense pressure on threatened vegetation types and allow diversity to increase, while at the same time not altering in any fundamental way the character of the New Forest system. Grazing pressure will continue to fluctuate for its own independent reasons; already increases in marking fees and a slackening market for ponies has resulted in a decline in animal numbers from their late 1970 levels (in 1984, marking fees were paid for only 3,000 ponies and 1,760 cattle). Such factors are outside our control: we can hope to manage only *within* the framework of a given animal density to 'spread' its effects in the most appropriate way. There will always be conflict between the interests of economic forestry, conservation, recreation and commoning. Perhaps the Forest is fortunate in that each interest has its own statutory body to champion its cause: Forestry Commission, Nature Conservancy Council, Countryside Commission and Verderers Court. The New Forest has survived the vagaries and vicissitudes of 900 years of changing use: it would be interesting to follow its fortunes over the next 900 years.

References

Adams, S.N. (1975) Sheep and cattle grazing in forests: a review. *Journal of applied Ecology 12*: 143-52.

Andrzejewska, L. (1967) Estimation of the effects of feeding of the sucking insect *Cicadella viridis* L. (Homoptera: Auchenorrhyncha) on plants. In K. Petrusewicz (ed.),*Secondary Productivity in Terrestrial Ecosystems*, vol. II: 791-805.

Archer, M. (1973) Variations in potash levels in pastures grazed by horses; a preliminary communication. *Equine Veterinary Journal 5*: 45-6.

—— (1978) Studies on producing and maintaining balanced pastures for horses. *Equine Veterinary Journal 10*: 54-9.

Arnold, G.W., and Dudzinski, M.L. (1978) *Ethology of free-ranging domestic animals. Developments in Animal and Veterinary Sciences, 2,* Elsevier.

Bailey, R.E. (1979) Use of faecal techniques in analysis of population density and diet in fallow deer populations. BSc Honours thesis (Environmental Sciences), University of Southampton.

—— and Putman, R.J. (1981). Estimation of fallow deer (*Dama dama*) populations from faecal accumulation. *Journal of applied Ecology 18*: 697-702.

Berger, J. (1977) Organisational systems and dominance in feral horses in the Grand Canyon. *Behavioural Ecology and Sociobiology 2*: 131-46.

de Bie, S. (1982) Carrying-capacity. Paper to meeting '*Begrazing door Vertebraten*', Rijksuniversiteit Groningen — Landbouwhogeschool, Wageningen. 1982.

Bobek, B., Perzanowski, K. Siwanowicz, J., and Zielinski, J. (1979) Deer pressure on forage in a deciduous forest. *Oikos 32*: 373-9.

Bonsma, J.C., and Le Roux, J.D. (1953) Influence of environment on the grazing habits of cattle. *Farming in South Africa 28*: 43-6.

Browning, D.R. (1951) The New Forest pastoral development scheme. *Agriculture 58*: 226-33.

Buechner, H.K., and Dawkins, H.C. (1961). Vegetation change induced by elephants and fire in Murchison Falls National Park, Uganda. *Ecology 42*: 752-66.

Bullen, F.T. (1970) A review of the assessment of crop losses caused by locusts and grasshoppers. *Proceedings of the International Study Conference on Current and Future Problems of Acridology*: 163-71. London.

Cadman, W.A. (1966) *The Fallow Deer.* Forestry Commission Leaflet No. 52. HMSO: London.

Carter, N.A. (1981) A study of bole damage by sika in England, with particular reference to bole-scoring. BSc thesis, University of Bradford.

Chadwick, M.J. (1960) *Nardus stricta* L. *Journal of Ecology 48*: 255-67.

Chapman, D.I., and Chapman, N. (1976) *The Fallow Deer.* Terence Dalton.

Chapman, N. (1984) *Fallow deer.* Mammal Society/Anthony Nelson.

Clements, R.O. (1978) The benefits and some long-term consequences of controlling invertebrates in a perennial rye grass sward. *Scientific Proceedings of the Royal Dublin Society, Series A, 6*: 335-41.

Clutton-Brock, T.H., and Albon, S.D. (1978) The roaring of red deer and the evolution of honest advertisement. *Behaviour 69*: 145-70.

—— (1979) Sexual differences in ecology between male and female red deer. Unpublished paper to the Ungulate Research Group Winter meeting, Edinburgh, December 1979.

Corke, D. (1974) The Comparative Ecology of the two British species of the genus *Apodemus* (Rodentia: Muridae). PhD thesis, University of London.

Countryside Commission (1984) *The New Forest Commoners.* Countryside Commission: Cheltenham.

Coupland, R.T. (1979) *Grassland Ecosystems of the World: Analysis of Grasslands and their Uses.* Cambridge University Press: Cambridge.

Crawley, M.J. (1983) *Herbivory: The Dynamics of Animal-Plant Interactions.* Blackwell Scientific Publications: Oxford.

Davidson, J.L., and Milthorpe, F.L. (1966) Leaf growth in *Dactylis glomerata* following defoliation. *Annals of Botany* (London) *30*: 173-84.

Davis, T.A.W. (1960) Kestrel pellets at a winter roost. *British Birds 53*: 281-4.

—— (1975) Food of the Kestrel in winter and early spring. *Bird Study 22*: 85-91.

Detling, J.K., Dyer, M.I., Proctor-Gregg, C., and Winn, D.T. (1980) Plant-herbivore interactions: Examination of potential effects of bison saliva on regrowth of *Bouteloua gracilis* (H.B.K.) Lag. *Oecologia 45*: 26-31.

Dinesman, L.G. (1967) Influence of vertebrates on primary production of terrestrial communities. In K. Petrusewicz (ed.) *Secondary Productivity in Terrestrial Ecosystems,* vol. 1: 261-6.

Duncan, P. (1983) Determinants of the use of habitat by horses in a Mediterranean wetland. *Journal of Animal Ecology 52*: 93-109.

—— and Vigne, N. (1979) The effect of group size in horses on the rate of attacks by blood-sucking flies. *Animal Behaviour 27*: 623-5.

Dzieciolowski, R. (1979) Structure and spatial organisation of deer populations. *Acta Thieriologica 24*: 3-21.

Edwards, P.J., and Hollis, S. (1982) The distribution of excreta on New Forest grasslands used by cattle, ponies and deer. *Journal of applied Ecology 19*: 953-64.

Ellis, J.E.S. (1946) Notes on the food of the Kestrel. *British Birds 39*: 113-15.

Ellison, L. (1960) Influence of grazing on plant succession of rangelands. *Botanical Review 26*: 1-78.

Estes, R.D. (1967) The comparative behaviour of Grant's and Thompson's gazelles. *Journal of Mammalogy 48*: 189-209.

Feist, J.D., and McCullough, D.R. (1976) Behaviour patterns and communication in feral horses. *Zeitschrift für Tierpsychologie 41*: 337-71.

Furubayashi, K., and Maruyama, N. (1977) Food habits of sika deer in Fudakake, Tanzawa Mountains. *Journal of the Mammal Society of Japan 7*: 55-62.

Gates, S. (1979) A study of the home ranges of free-ranging Exmoor ponies. Mammal Review 9: 3-18.

—— (1982) The Exmoor Pony — a wild animal? *Nature in Devon 2*: 7-30.

Gessaman, J.A., and MacMahon, J.A. (1984) Mammals in ecosystems: their effects on the composition and production of vegetation. *Acta Zoologica Fennica,* in press.

Gill, E.L. (1980) Some aspects of the social and reproductive behaviour of a group of New Forest ponies. BSc Honours thesis (Environmental Sciences), University of Southampton.

—— (1984) Seasonal changes in body condition of New Forest ponies. Internal report, University of Southampton.

Grant, S.A., and Hunter, R.F. (1966) The effects of frequency and season of clipping on the morphology, productivity and chemical composition of *Calluna vulgaris.* *New Phytologist 65*: 125-33.

Gray, A.J., and Scott, R. (1977) *Puccinellia maritima* (Huds.) Par 1. *Journal of Ecology 65*: 699-716.

Hansen, R. (1970) Foods of free-roaming horses in Southern New Mexico. *Journal of Range Management 29*: 347.

Harper, J.L. (1977) *Population Biology of Plants.* Academic Press: London.

Harrington, R. (1981) Immuno-electrophoresis and the genetics of red/sika hybrids. Unpublished paper given to Ungulate Research Group Winter meeting, Southampton, December 1981.

Hill, S.D. (1985) Influences of large herbivores on small rodents in the New Forest, Hampshire. PhD thesis, University of Southampton.

Hirons, G.J.M. (1976) A population study of the Tawny Owl (*Strix aluco*) and its main prey species in woodland. DPhil thesis, University of Oxford.

—— (1984) The diet of Tawny Owls (*Strix aluco*) and Kestrels (*Falco tinnunculus*) in the New Forest, Hampshire. *Proceedings of the Hampshire Field Club and Archaeological Society 40*: 21-6.

Hoffmeyer, I. (1973) Interaction and habitat selection in the mice *Apodemus flavicollis* and *Apodemus sylvaticus. Oikos 24*: 108-16.

Hofmann, R.R. (1982) Morphological classification of sika deer within the comparative system of ruminant feeding types. *Deer 5*: 252-3.

—— and Stewart, D.R.M. (1972) Grazer or browser? A classification based on the stomach structure and feeding habits of East African ruminants. *Mammalia 36*: 226-40.

—— Geiger, G., and Konig, R. (1976) Vergleichend-anatomische untersuchungen an der vormagenschleimhaut von Rehwild (*Capreolus capreolus*) und Rotwild (*Cervus elaphus*). *Zeitschrift Säugetierkunde 41*: 167-93.

Hope Simpson, J.F. (1940) Studies of the vegetation of the English Chalk. VI. Late stages in succession leading to chalk grassland. *Journal of Ecology 28*: 386-402.

Horwood, M.T. (1973) The world of the Wareham sika. *Deer 2*: 978-84.

—— and Masters, E.H. (1970) *Sika Deer.* British Deer Society: Reading.

Hosey, G.R. (1974) The food and feeding ecology of roe deer. PhD thesis, University of Manchester.

Howard, P.C. (1979) Variability of feeding behaviour in New Forest ponies. BSc Honours thesis (Biology), University of Southampton.

Jackson, J.E. (1974) The feeding ecology of Fallow deer in the New Forest, Hampshire. PhD thesis, University of Southampton.

—— (1977) The annual diet of the Fallow deer (*Dama dama*) in the New Forest, Hampshire, as determined by rumen content analysis. *Journal of Zoology* (London) *181*: 465-73.

—— (1980) The annual diet of the roe deer (*Capreolus capreolus*) in the New Forest, Hampshire, as determined by rumen content analysis. *Journal of Zoology* (London) *192*: 71-83.

Jameson, D.A. (1963) Responses of individual plants to harvesting. *Botanical Review 29*: 532-94.

Jarman, P.J. (1974) The social organisation of antelope in relation to their ecology. *Behaviour 48*: 215-66.

Johnson, T.H. (1984) Habitat and social organisation of roe deer (*Capreolus capreolus*). PhD thesis, University of Southampton.

Jones, M.G. (1933) Grassland management and its influence on the sward. *Journal of the Royal Agricultural Society 94*: 21-41.

Kaluzinski, J. (1982) Composition of the food of roe deer living in fields and the effects of their feeding on plant production. *Acta Theriologica 27*: 457-70.

Kay, R.N.B. (1979) Seasonal changes of appetite in deer and sheep. Agricultural Research Council. Research Review 1979.

—— and Suttie, J.M. (1980) Seasonal cycles of voluntary food intake in red deer. Unpublished paper presented to the Ungulate Research Group winter meeting, Cambridge, December 1980.

Kenchington, F.E. (1944) *The Commoners' New Forest.* Hutchinson: London.

Klingel, H. (1974) A comparison of the social behaviour of the Equidae. In V. Geist and F. Walther (eds.), *The Behaviour of Ungulates and its relation to Management.*

Krefting, L.W., Stenlund, M.H., and Seemel, R.K. (1966) Effect of simulated and natural browsing on mountain maple. *Journal of Wildlife Management 30*: 481-8.

Kruuk, H. (1984) Group sizes and home ranges of carnivores. In R.H. Smith and R.M. Sibly (eds.), *Behavioural Ecology: the ecological consequences of adaptive behaviour.* British Ecological Society Symposium 25.

Lascelles, G.W. (1915) *Thirty five years in the New Forest.* London.

Laws, R.M. (1976) Elephants as agents of habitat and landscape change in East Africa. *Oikos 21*: 1-15.

Lincoln, G.A., and Guinness, F.E. (1973). The sexual significance of the rut in red deer. *Journal of Reproduction and Fertility,* Supplement *19*: 475-89.

—— Youngson, R.W., and Short, R.V. (1970) The social and sexual behaviour of the red deer stag. *Journal of Reproduction and Fertility,* Supplement *11*: 71-103.

Lowe, V.P.W., and Gardiner, A.S. (1975) Hybridisation between red deer and sika deer, with reference to stocks in North-west England. *Journal of Zoology* (London) *177*: 553-66.

MacArthur, R.H., and Levins, R. (1967). The limiting similarity, convergence and divergence of coexisting species. *American Naturalist 101*: 377-85.

Mann, J.C.E. (1978) An investigation into the vegetational differences between an ungrazed and a grazed portion of the New Forest, with special emphasis on the species composition, species diversity and productivity of the areas. BSc Honours thesis (Biology), University of Southampton.

—— (1983) The Social Organisation and Ecology of the Japanese Sika deer (*Cervus nippon*) in Southern England. PhD thesis, University of Southampton.

McEwen, L.C., French, C.E., Magruder, N.D., Swift, R.W., and Ingram, R.H. (1957) Nutrient requirements of the white-tailed deer. *Transactions of the 22nd North American Wildlife Conference*: 119-32.

McNaughton, S.J. (1979a) Grazing as an optimisation process: grass-ungulate relationships in the Serengeti. *American Naturalist 113*: 691-703.

—— (1979b) Grassland-herbivore dynamics. In A.R.E. Sinclair and M. Norton-Griffiths (eds.), *Serengeti: dynamics of an ecosystem.* University of Chicago Press, 46-81.

Milthorpe, F.L., and Davidson, J.L. (1965) Physiological aspects of regrowth in grasses. In F.L. Milthorpe and J.D. Ivins (eds.), *The Growth of Cereals and Grasses.* Butterworth: London, 241-55.

Miura, S. (1974) On the seasonal movements of sika deer populations in Mt. Hinokiboramu. *Journal of the Mammal Society of Japan 6*: 51-66.

Nicholson, I.A., Paterson, I.S., and Currie, A. (1970) A study of vegetational dynamics: selection by sheep and cattle in *Nardus* pasture. In A. Watson (ed.), *Animal Populations in Relation to their Food Resources.* BES Symposium *10*: 129-43.

O'Bryan, M.K., Palmes, P., and Putman, R.J. (1980) Factors affecting group size and cohesiveness in populations of New Forest ponies. Unpublished MS, Southampton University Library.

Odberg, F.O., and Francis-Smith, K. (1976) A study on eliminative and grazing behaviour: the use of the field by captive horses. *Equine veterinary Journal 8*: 147-9.

—— (1977) Studies on the formation of ungrazed eliminative areas in fields used by horses. *Applied Animal Ethology 3*: 27-34.

Oosterveld, P. (1981) Begrazing als natuurtechnische maatregel. Unpubl. ms.

Rijkinstituut voor Natuurbeheer, The Netherlands.

Packham, C.G. (1983) The influence of food supply on the ecology of the badger. BSc Honours thesis (Biology), University of Southampton.

Parfitt, A. (1985) 'Social Organisation and Ecology of fallow deer'(*Dama dama*) in the New Forest, Hampshire. PhD thesis, University of Southampton, in preparation.

Parsons, K.A., and De La Cruz, A.A. (1980) Energy flow and grazing behaviour of conocephaline grasshoppers in a *Juncus roemerianus* marsh. *Ecology 61*: 1045-50.

Peterken, G.F., and Tubbs, C.R. (1965) Woodland regeneration in the New Forest, Hampshire since 1650. *Journal of applied Ecology 2*: 159-70.

Pianka, E.R. (1973) The structure of lizard communities. *Annual Review of Ecology and Systematics 4*: 53-74.

Pickering, D.W. (1968) Heathland reclamation in the New Forest: the ecological consequences. Unpublished MSc thesis, University College, London.

Pollock, J.I. (1980) *Behavioural Ecology and Body Condition Changes in New Forest Ponies*. Royal Society for the Prevention of Cruelty to Animals. Scientific Publications No. 6.

Pratt, R.M., Putman, R.J., Ekins, J.R., and Edwards, P.J. (1985) Use of habitat by free-ranging cattle and ponies in the New Forest, Southern England. In press.

Prior, R. (1973) Roe deer: management and stalking. Game Conservancy Booklet No. 17, The Game Conservancy: Fordingbridge.

Prisyazhnyuk, V.E., and Prisyazhnyuk, N.P. (1974) [Sika deer on Askold Island.] Bulletin Moskow o-va ispyt. Priv. otd. Biology *79*: 16-27 (in Russian).

Putman, R.J. (1980) Consumption, protein and energy intake of fallow deer fawns on diets of differing nutritional quality. *Acta Theriologica 25*: 403-13.

—— (1981) Social systems in deer: a speculative review. *Deer 5*: 186-8.

—— (1983) Nutrition of Wild and Farmed Fallow deer. Paper to 1st International Conference on the Biology of Deer Production, Dunedin, New Zealand.

—— (1984) Facts from Faeces. *Mammal Review 14*: 79-97.

—— (1985a) Effects of grazing mammals on ecosystem structure and function: a review. In press.

—— (1985b) Competition and Coexistence in a multispecies grazing system, *Acta Theriologica*, in press

—— and Wratten, S.D. (1984) *Principles of Ecology*. Croom Helm: Beckenham.

—— Edwards, P.J., Ekins, J.R., and Pratt, R.M. (1981) Food and feeding behaviour of cattle and ponies in the New Forest: a study of the inter-relationship between the large herbivores of the Forest and their vegetational environment. Report HF 3/03/127 to Nature Conservancy Council: Huntingdon.

—— Pratt, R.M., Ekins, J.R., and Edwards, P.J. (1982) Habitat use and grazing by free-ranging cattle and ponies, and impact upon vegetation in the New Forest Hampshire. Proceedings of IIIrd International Theriological Congress, Helsinki. *Acta Zoologica Fennica, 172*, 183-6

—— (1985) Food and feeding behaviour of cattle and ponies in the New Forest, Hampshire. (in prep.)

Rand, N. (1981) Fallow deer coat patterns: a system of identification and some results from its use in a long-term study of behaviour. Unpublished paper given to Ungulate Research Group Winter meeting, Southampton, December 1981.

Rawes, M. (1981) Further results of excluding sheep from high level grasslands in the north Pennines. *Journal of Ecology 69*: 651-69.

Reardon, P.O., Leinweber, C.L., and Merrill, L.B. (1972) The effect of bovine saliva on grasses. *Journal of Animal Science 34*: 897-8.

—— (1974) Response of sideoats grama to animal saliva and theamine. *Journal of Range Management 27*: 400-1.

Russell, V. (1976) *New Forest Ponies*. David and Charles: Newton Abbott.

Schmidt, P.J. (1969) Observations on the behaviour of cattle in a hot dry region of the Northern Territory of Australia, with particular reference to walking, watering and grazing. MSc thesis, The University of New England, Armidale, New South Wales.

Schröpfer, R. (1983) The effect of habitat selection on distribution and spreading in the yellow-necked mouse (*Apodemus flavicollis*). Proceedings of IIIrd International Theriological Congress, Helsinki. *Acta Zoologica Fennica.*

Schultz, A.M. (1964) The nutrient-recovery hypothesis for arctic microtine cycles. In D.J. Crisp (ed.), *Grazing in Terrestrial and Marine Environments.* Blackwell: Oxford, 57-68.

—— (1969) A study of an ecosystem: the arctic tundra. In G.M. van Dyne (ed.), *The Ecosystem Concept in Natural Resource Management.* Academic Press: New York, 77-93.

Schuster, M.F., Boling, J.C., and Morony, J.J. (1971) Biological control of rhodesgrass scale by airplane release of an introduced parasite of limited dispersal ability. In C.B. Huffaker (ed.), *Biological Control.* Plenum Press: New York, 227-50.

Shepperd, J.H. (1921) The trail of the short grass steer. North Dakota Agricultural College Bulletin *154*: 8.

Short, H.L., Newson, J.D., McCoy, G.L., and Fowler, J.F. (1969) Effects of nutrition and climate on southern deer. *Transactions of the 34th North American Wildlife Conference*: 137-45.

Simenstad, C.A., Estes, J.A., and Kenyon, K.W. (1978) Aleuts, sea-otters, and alternative stable-state communities. *Science 200*: 403-10.

Sinclair, A.R.E., and Norton-Griffiths, M. (1979) *Serengeti: Dynamics of an Ecosystem.* University of Chicago Press: Chicago.

Smith, R.H. (1979) On selection for inbreeding in polygynous animals. *Heredity 43*: 205-11.

Southern, H.N. (1954) Tawny owls and their prey. *Ibis 98*: 384-410.

—— (1969) Prey taken by Tawny owls during the breeding season. *Ibis 111*: 293-9.

—— (1970) The natural control of a population of Tawny Owls (*Strix aluco*). *Journal of Zoology* (London) *162*: 197-285.

Spedding, C.R.W. (1971) *Grassland Ecology.* Clarendon Press: Oxford.

Spence, D.H.N., and Angus, A. (1971) African grassland management: burning and grazing in Murchison Falls National Park, Uganda. In E. Duffey and K. Watt (eds.), *The Scientific Management of Animal and Plant Communities for Conservation.* Blackwell: Oxford, 319-31.

Staines, B.W. (1976) Experiments with rumen-canulated red deer to evaluate rumen analyses. *Journal of Wildlife Management 40*: 371-3.

—— Crisp, J.M., and Parish, T. (1981) Differences in the quality of food eaten by red deer stags and hinds in winter. *Journal of applied Ecology 19*: 65-79.

Stebbins, G.L. (1981) Coevolution of grasses and herbivores. *Annals of the Missouri Botanical Garden 68*: 75-86.

Stewart, D.R.M. (1967) Analysis of plant epidermis in faeces: a technique for studying the food preferences of grazing herbivores. *Journal of applied Ecology 4*: 83-111.

Stewart, D.R.M., and Stewart, J. (1970) Food preference data by faecal analysis for African plains ungulates. *Zoological Africana 5*: 115.

Strange, M.L. (1976) A computer study of the fallow deer of the New Forest. BSc Honours thesis (Environmental Sciences), University of Southampton.

Takatsuki, S. (1980) Food habits of sika deer on Kinkazan Island, Japan. *Scientific Reports of Tohaku University, Series 4* (Biology), *38*: 7-32.

Tansley, A.R. (1922) Studies of the vegetation of the English Chalk. II. Early stages of redevelopment of woody vegetation in chalk grassland. *Journal of Ecology 10*: 168-77.

—— and Adamson, R.W. (1925) Studies of the vegetation of the English Chalk. III. The chalk grasslands of the Hampshire-Sussex border. *Journal of Ecology 13*: 177-223.

Tubbs, C.R. (1965) The development of the small-holding and cottage stock-keeping economy of the New Forest. *Agricultural Historical Review 13.*

—— (1968) *The New Forest: an ecological history.* David and Charles: Newton Abbott.

—— (1974) *The Buzzard.* David and Charles: Newton Abbot.

—— (1982) The New Forest: conflict and symbiosis. *New Scientist* 1st July 1982: 10-13.

—— and Jones, E.L. (1964) The distribution of gorse (*Ulex europaeus*) on the New Forest in relation to former land use. *Proceedings of the Hampshire Field Club 23:* 1-10.

—— and Tubbs, J.M. (1985) Buzzards, *Buteo buteo,* and land use in the New Forest, Hampshire, England. *Biological Conservation 31:* 41-65.

Turner, D.C. (1979) An analysis of time-budgetting by roe deer (*Capreolus capreolus*) in an agricultural area. *Behaviour 71:* 246-90.

Tyler, S. (1972) The behaviour and social organisation of the New Forest ponies. *Animal Behaviour Monographs 5:* 87-194.

van der Veen, H.E. (1979) Food selection and habitat use in the red deer (*Cervus elaphus* L.). PhD thesis, Rijksuniversiteit te Groningen.

van Wieren, S.E. (1979) Sexual segregation in deer species: a possible theory. Unpublished paper to the Ungulate Research Group Winter meeting Edinburgh, December 1979.

Vickery, P.J. (1972) Grazing and net primary production of a temperate grassland. *Journal of applied Ecology 9:* 307-14.

Vittoria, A., and Rendina, N. (1960) Fattori condizionanti la funzionalita tiaminica in piante superiori e cenni sugli effetti dell bocca del ruminanti sull erbe pascolative. *Acta Medica Veterinaria (Naples), 6:* 379-405.

Wells, S. and von Goldschmidt-Rothschild, B. (1979) Social behaviour and relationships in a herd of Camargue horses. *Zeitschrift für Tierpsychologie 49:* 363-80.

Welsh, D. (1975) Population, behavioural and grazing ecology of the horses of Sable Island, Nova Scotia. PhD thesis, Dalhousie University.

Whittaker, R.H., and Woodwell, G.M. (1968) Dimension and production relations of trees and shrubs in the Brookhaven Forest, New York. *Journal of Ecology 56:* 1-25.

Wiegert, R.G., and Evans, F.C. (1967) Investigations of secondary productivity in grasslands. In K. Petrusewicz (ed.), *Secondary Productivity in Terrestrial Ecosystems,* vol. II: 499-518.

Wolff, J.O. (1978) Burning and browsing effects on willow growth in interior Alaska. *Journal of Wildlife Management 42:* 135-40.

Yalden, D.W., and Warburton, A. (1979) A diet of the Kestrel in the Lake District. *Bird Study 26:* 163-70.

Zejda, J. (1978) Field grouping of roe deer (*Capreolus capreolus*) in a lowland region. *Folia Zoologica 27:* 111-22.

Zhuopu, G., Enyu, C., and Youzhin, W. (1978) A new subspecies of sika deer from Szechuan (*Cervus nippon sichuanensis*). *Acta Zoologica Sinica 24:* 187-92.

Index